COUNTERSPACE

THE NEXT HOURS OF
WORLD WAR III

Forge Books by
William B. Scott, Michael J. Coumatos, and William J. Birnes

Space Wars

COUNTERSPACE

THE NEXT HOURS OF
WORLD WAR III

WILLIAM B. SCOTT
MICHAEL J. COUMATOS
WILLIAM J. BIRNES

A TOM DOHERTY ASSOCIATES BOOK

NEW YORK

Although the characters and actual events depicted in *Counterspace* are fiction and bear no relationship to actual persons living or dead, the action and weapons depicted herein are based on actual technologies and war-gaming strategies used by the United States military and civilian war planners in preparation for the types of events depicted fictitiously in *Counterspace*.

COUNTERSPACE: THE NEXT HOURS OF WORLD WAR III

A Forge Book
Published by Tom Doherty Associates, LLC
175 Fifth Avenue
New York, NY 10010

www.tor-forge.com

Forge® is a registered trademark of Tom Doherty Associates, LLC.

Library of Congress Cataloging-in-Publication Data

Scott, Bill, 1946–
   Counterspace : the next hours of World War III / William B. Scott,
Michael J. Coumatos, and William J. Birnes.—1st hardcover ed.
       p. cm.
   "A Tom Doherty Associates book."
   ISBN 978-0-7653-2232-6
   1. Satellites—Fiction.   2. International relations—Fiction.   3. War
games—Fiction.   4. World War III—Fiction.   I. Coumatos, Michael J.
II. Birnes, William J.   III. Title.
   PS3619.C655C68 2009
   813'.6—dc22

                                                          2009017173

First Edition: October 2009

Printed in the United States of America

0 9 8 7 6 5 4 3 2 1

Dedicated to
America's military men and women,
past and present

ACKNOWLEDGMENTS

The authors acknowledge our editors, Bob Gleason and Eric Raab, for their kindness and expert, insightful recommendations, and our publisher, Tom Doherty, for his courage in publishing our second work about combat in space. Our sincere thanks also go to a dynamic Tor / Forge partner, Karen Lovell, for her professionalism and tireless dedication in promoting *Space Wars: The First Six Hours of World War III*, our first book in this series. Our deep appreciation and thanks go to the myriad experts and gracious supporters who provided valuable suggestions, insights, and technical information. They include: Major General Erika Steuterman; Colonel Guy "Spike" Morley; Colonel Jim Wallace; Dr. Michael Colvard; Lieutenant Colonel Pete Garretson; Major Tod Pingrey (Ret.); Roz Brown; The Space Foundation; Phil Brooks; Tom Menza; and Mike Beuster. Profound "thank-yous" also go to our life partners, Linda Scott, Susan Coumatos, and Nancy Hayfield, who provided honest criticism and heartfelt support.

COUNTERSPACE

THE NEXT HOURS OF
WORLD WAR III

1 MAYHEM IN THE HEAVENS

5 MARCH 2011/GEOSYNCHRONOUS ORBIT

An aging Defense Support Program (DSP) satellite positioned 22,500 miles above the western Pacific Ocean was the first to spot a white-hot tail of fire streaking skyward from the Korean peninsula. The spacecraft's infrared telescope captured the missile's heat signature, processed the raw data, and passed them to a microprocessor. That chip compressed millions of digital bits and shot them at light speed to a powerful transmitter, which fired a stream of encrypted pulses to golf-ball-like domes sheltering massive antennas on Earth. Microseconds later, the infrared information appeared as speckles of light on a screen inside the windowless DSP Mission Control center at Buckley Air Force Base, near Denver.

An Air Force staff sergeant operator noted the DSP-detected IR signature, tagged it as an unusual item of interest, and confirmed the information was already triggering automated alerts at the North American Aerospace Defense Command (NORAD) Missile Warning Center at Peterson Air Force Base, roughly 80 miles to the south. There, alarms were going off, triggering a well-rehearsed flurry of activity.

The NORAD Command Director cross-checked the data and

quickly determined this was, indeed, a missile launch from North Korea. Radars in the Pacific had yet to determine a trajectory, but the required dual-phenomenology criteria were satisfied: Both the DSP's infrared sensor and ground-based radar data said this was real. A potentially deadly missile was en route—somewhere. *The crazy little bastard did it!* the officer thought grimly.

The NORAD director immediately initiated a missile-event tele-conference, knowing the President of the United States and the Canadian Prime Minister would be on the line very soon. *Twice in a matter of months! What the hell is going on?* the colonel wondered, throat tightening. He'd been on duty the previous year, when Iran launched a nuclear-tipped Shahab-4 missile at a U.S. air base in Italy. Fortunately, degraded navigation signals from maser-damaged Global Positioning System satellites had misguided that ballistic missile to a watery grave, and one of Iran's few, highly prized nuclear warheads had fizzled instead of going high-order.

The colonel's role was to immediately determine whether North America was under attack, then issue a simple yea or nay. So far, his people couldn't tell; the missile was still streaking straight up, its liquid-fueled engines relentlessly delivering thousands of pounds of thrust. He relayed only what he *did* know: A long-range Taepodong-4 missile had launched from a North Korean site that NORAD, U.S. Strategic Command (STRATCOM), and most of America's spook agencies had been monitoring constantly for several days.

A recent flurry of activity was detected by round-the-clock National Reconnaissance Office "spy" guardians orbiting in deep space, attracting considerable attention. Consequently, the North's launch came as no surprise. But where the missile was headed, and what it was carrying, might be. And these days, the United States' national security complex loathed surprises originating in Asia.

The powerful missile easily could be armed with one of North Korea's few nuclear weapons, a primitive atomic device that most assuredly had been rat-holed in deep-underground bunkers since late 2007, when the hermetic nation had agreed to dismantle its nuclear program. None of the Central Intelligence Agency's Eastern Asia experts believed Kim Jong Il had actually accounted for all of his fissile material, even though international inspection agencies declared the North "nuclear free," except for Western-built nuclear reactors generating electrical power. The CIA knew quite well the crazy-like-a-fox dictator had hidden several nukes. Where they were hidden, the CIA didn't know, but obviously far from the prying eyes of those international experts who *so* wanted to believe that North Korea had truly abandoned its nuclear weapons program.

"Where are you headed, Dong-bird . . . ?" the director muttered, his eyes riveted to huge wall-mounted screens displaying a computer-generated track of the Taepodong's flight path.

"Missile launch confirmed. Type is a North Korean Taepodong-Four. Trajectory undetermined. For now, North America is not—repeat *not*—under attack," he finally said. His preliminary assessment was instantly routed to Washington, Ottawa, and several key military command centers scattered across the North American continent.

GEOSYNCHRONOUS ORBIT/SPACE-BASED INFRARED SYSTEM

Cued by the DSP spacecraft's infrared detector, one of the U.S.'s new Space Based Infrared System, or SBIRS, platforms was also tracking the rapid ascent of that Taepodong missile. The satellite's sensitive infrared staring array and sophisticated tracking algorithms were poised, ready to detect any hint of the missile starting to arc over, which would reveal its ultimate target. Once that arc was identified, computers would define an oblong "footprint," projecting where the missile and its ominous warhead might impact.

But it hadn't pitched over yet. Instead, the missile continued to climb vertically, streaking through the atmosphere's thin upper layer and into the almost-vacuum of near-space.

One hundred miles . . . 120 miles . . . 130 . . . Then a massive explosion of scorching ultra-white light temporarily blinded both the DSP and SBIRS spacecraft's thermally sensitive "eyes." Instantly, nuclear detectors, or nudets, on both of the missile-warning platforms, plus similar nudets on a handful of Global Positioning System (GPS) satellites in mid-Earth orbit, registered a barrage of radioactive particles and high-energy gamma rays. Once the mayhem was sorted out, all those nudet alarms were converted to electronic signals beamed back to Earth.

The NORAD Command Director choked on his coffee, unable to fully grasp what he was witnessing. *Holy shit! Nudet! The dumb shits . . .* His eyes scanned several screens. Other officers in the center were confirming what he saw, their reports delivered in crisp, concise terms. He squeezed a phone receiver against his ear, voicing words he'd hoped would never cross his lips, not yet comprehending their ramifications.

"Nudet! We have a nuclear detonation . . . in space! No altitude reading at this time. Repeat: A confirmed nudet was recorded, while the Taepodong missile was still on ascent." The director rapidly scanned the oversized wall display, as well as several computer screens at the command position. His mind raced ahead, while also sifting through the professional summaries from his crew. NORAD missile-warning specialists' concise, unemotional reports rippled across the command center, following procedures that had been practiced for decades. A message popped up in the corner of a computer screen, drawing the director's attention. *There!* A key confirmation he'd been looking for, a critical assessment from the Air Force's top-dog nuclear experts: "AFTAC confirms nudet at 137-mile altitude," it read.

Sensitive nuclear detectors mounted on DSP and SBIRS missile-warning satellites, as well as backups on GPS navigation spacecraft, had registered massive emissions of deadly gamma rays and beta particles—high-energy electrons—that could only have come from a nuclear detonation. Those data were funneled from stations scattered across the globe to an operations post in the Air Force Technical Applications Center headquarters at Patrick Air Force Base, Florida. Once a super-secret unit, AFTAC was the clearinghouse for all data acquired by space-based nuclear detectors. Its highly trained experts would soon determine precisely how powerful and sophisticated the North Korean weapon had been.

Not since the tense Cold War days of the 1960s had a nuclear weapon exploded in or on the edge of space. Today, the world's first space war had just reignited in dramatic fashion.

U.S. STRATEGIC COMMAND HEADQUARTERS/ COMMAND POST

General Howard Aster's lengthy strides forced his aide, Lieutenant Colonel Thad "Burner" Burns, who also served as U.S. Strategic Command's executive officer, to break into a trot. Aster swiped a coded badge through a cipher-lock reader beside the command post's heavy door and entered his security code. A solid snap announced the STRATCOM commander was authorized entry. Aster grabbed the door's handle and flung the barrier wide, racing into the room and leaping a few steps up to the battle cab.

"What do we have, Dave? Where's it headed?" Aster demanded. The four-star former fighter pilot had left the command's wargame center the instant his handheld communicator had sounded off and displayed a chilling phrase on its tiny screen: NK MSL LAUNCH.

"Sir, the North Koreans just detonated a nuke in space, about a hundred and thirty-seven miles high," Army Lieutenant General Dave Forester clipped, eyes still scanning the command post's wall-size screens. Graphics and text flashed across a single giant display dedicated to the Pacific Command, or PACOM, region.

"In space? What the hell . . .?" Aster's brow knitted, trying to make sense of this unexpected twist. The North Koreans had supposedly "denuclearized" four years ago, in compliance with a six-nation agreement that had sent tons of fuel oil and food to the starving nation since then. Evidently, the North's weird, but wily, dictator had squirreled away at least one nuclear device, successfully hiding the weapon from dozens of diligent inspectors. Diplomats from Washington to Tokyo, Seoul, and Beijing were still congratulating themselves on the success of those nuclear-disarmament negotiations, despite evidence that the North had restarted its weapons program.

So much for denuking-through-diplomacy, Aster thought, his rugged, square-jawed features reflecting the strain of recent months. Since America's satellites had come under attack in April 2010, crippling the United States' ability to monitor the activities of Earth's bad actors from orbiting platforms in space, Aster's STRAT-COM team had been under intense, round-the-clock pressure. As the nation's primary military center of strategic expertise, STRAT-COM assimilated intelligence and other data received from myriad sources across the globe, converted it to knowledge, then delivered options to top officials in Washington. Ultimately, Aster's job was to defend the United States through the employment of its long-reach military forces, but the first order of business was understanding any particular situation. At this moment, the general didn't.

Forester, STRATCOM's three-star director of operations, pointed to a dotted line on the big screen, its tail anchored in a graphic of the Korean peninsula. "We expected the missile to arc over at any second, but the damned thing blew up! Nudets on several space-craft registered prompt gamma and beta radiation, and the infrared missile-warning platforms—DSP and SBIRS—confirmed a super-hot explosion. Radars showed the missile had disappeared; no arc-over or trajectory whatsoever. No question, sir. The damned North Koreans popped a nuke on the fringe of space. Several of our space platforms and a hell of a lot of electronics stuff all over the Pacific region are probably fried."

Aster's mind was racing, trying to fully grasp the far-reaching im-plications of Forester's abbreviated assessment. "Get a status on all our spacecraft in LEO, Dave. Everything—national-security, civil, and commercial birds. If that thing went off at a hundred thirty-seven miles up, you're right; it created one hell of an electromag-netic pulse." A nuclear explosion 30–200 miles above the Earth's surface, in that broad region also known as Low-Earth orbit, or LEO, where the atmosphere thins and space begins, would produce a mas-sive flux of energetic electrons (beta particles) and create a powerful electromagnetic pulse. That EMP *would* induce unwanted current spikes in wires and circuits, frying sensitive electronics for hun-dreds of miles in all directions.

Forester fired a verbal order to a nearby officer, an Air Force ma-jor, who nodded and went to work at a large-screen workstation. Text flowed across his display, fed into the command center from hundreds of stations around the globe.

"Sir, prelim only," the major announced, reading the rapidly changing data, "but it looks like a DMSP, a couple of NRO imagers, and several sigint birds are down already. We're polling commercial operators right now, but a few Excalibur comsats and at least one EarthView imaging platform were in the potential-kill circle. Uh . . . Schriever's now showing some degradation of more GPS navsats, too, but not from EMP, of course. Gotta be from prompt radiation."

Aster mentally translated. Loss of a DMSP—a Defense Meteorological Satellite Program bird—meant military forces in the northern Pacific would be starving for timely weather data. A handicap, but not devastating, as long as those two Navy aircraft carrier groups steaming northward in the Pacific didn't run into a stretch of bad weather.

The National Reconnaissance Office's low-altitude electro-optical and signal-intelligence platforms, America's ultra-secret spy satellites, had taken a severe hit, though. Even the advanced constellation of small, quick-reaction surveillance "nanosats" that Brigadier General "Speed" Griffin had flown to orbit in the ultra-secret "Blackstar" XOV spaceplane almost a year ago more than likely were silenced. As quick-reaction spacecraft, the nanosats were built with commercial electronics gear, instead of military-specified "radiation-hardened" components. The other milspace birds were rad-hardened, but age and constant bombardment by cosmic rays had taken a toll on their resistance. Collectively, those spacecraft losses would definitely hurt intel-gathering in Pacific Command's Area of Responsibility, drastically reducing America's knowledge of activities inside North Korea, China, and other Pacific Rim locales. In essence, the nation's space-based eyes on that part of the globe were now blinded.

Shit! Even less situational awareness! Aster, the tall, trim, prematurely white-haired STRATCOM commander fumed silently, his jaw muscles twitching as he scanned myriad data presentations. *In sports terms*, the six-foot-seven former Air Force Academy basketball star mused grimly, *we have a terrific first team of national security satellites, but there's absolutely no bench. Our first string's been whacked severely, and we don't have replacements to send in. That could mean losing the whole damned game!*

The high-level cram course on nuclear weapons effects Aster had received, when he took over as STRATCOM chief, had provided a basic understanding of electromagnetic pulse effects. But that was thin knowledge, at best. The general needed more to assess the full spectrum of impacts now.

"Major, why would GPS be affected by that nuke? All the Navstars are in much higher mid-Earth orbits, something like twelve-thousand-plus miles, right? That's too high for EMP effects, 'cause there's no air to ionize that far out."

"Correct, sir. But that detonation created an extremely high radiation flux, and it's basically charging up the Van Allen Belt, even though *it's* way out there, too. In turn, that causes what we call 'secondary radiation effects' in electronic circuits on GPS birds—things like electronic gate latch-ups, data losses, and other effects. It also created an ion-charged layer in the upper atmosphere, which acts like a shield that blocks the weak signals from GPS and other satellites. Most GPS navigation and timing signals are now prevented from reaching Earth, especially out in the Pacific. Probably going to affect the downlinks from some GEO birds, too," the major added, referring to platforms in geostationary orbit, 22,500 miles from Earth.

Aster was impressed by the officer's concise, no-nonsense explanation. No techno-gibberish or lofty science to impress and baffle an old fighter-jock general. He glanced at a name tag on the major's flight suit, committing it to memory.

"Rog. Thanks," Aster nodded. Turning back to Forester, his STRATCOM operations chief, who had moved across the room, he called, "Dave! Anything from PACOM yet? What's the air and ground impact out there?" Aster tried to make sense of disparate data filling wall-size flat-panel screens ringing the command post, while Forester fielded documents from several officers scurrying about the battle cab in a hushed, yet professional ballet. The director of operations also kept glancing at the screens, trying to correlate a flood of information.

"Comm with our bombers is spotty, sir," Forester began. "We scrambled B-2s and some B-1s as soon as that Taepodong launched, just to get air assets clear of potential North Korean targets. The bombers are holding at predesignated locations, but PACOM's having a hell of a time maintaining contact with 'em," Forester said. "Got some electrical power outages across Japan, South Korea, and Okinawa. TV stations off the air . . . Phone systems—both landline and cell networks—are a shambles everywhere. Guam seems to be in pretty good shape, though. Hawaii is definitely outside the kill-ring, too. Looks like the biggest hit to us was our LEO spacecraft. India and Japan also had some exposed birds in LEO, but no word on what happened to them yet."

Aster was nodding, realization dawning first, followed quickly by conviction. "The sumbitches were trying to do exactly that! Knock out everything in space that could get a good look at 'em. And wipe out our ability to communicate efficiently all across that area of the Pacific. Hell! You can bet that's not the full story, either!" The general turned back to the nuclear-smart major, clasping his shoulder as the officer continued to scan his workstation's oversized screen, now blanketed with numerous open windows.

"Major! An EMP from that altitude . . . What would it do to 'tronics on the ground in both South *and* North Korea?" Aster demanded.

Startled, the officer recoiled, but responded readily. "Sir, the entire power grid in both countries are probably knocked out, because control system microcircuits would have been zapped. Radios would be dead. Computer chips in everything from phones to car ignition systems to commercial PCs could be fried. About anything electrical that wasn't specifically hardened against EMP would be wiped out."

Aster nodded, thanking the officer absently. *The dumbshits just blinded themselves, too! They'd never see our bombers coming— Hell! They don't think we'll retaliate!* He was starting to understand the North Korean leader's wily strategy. Or at least part of it.

"Dave, is PACOM in contact with those two carrier groups? One was headed for the South China Sea and the other's in the Sea of Japan, last I checked."

Forester queried another officer, before responding. "Intermittently, sir. We can still use a few sat-birds to reach 'em, but Milstar's our primary link, and its downlink channels are flaky, for some reason. And we've lost the unmanned, near-space platform shadowing those carriers. Looks like the nuke disabled its entire comm-relay electronics package; it's toast. For comm relay, that HARVe's (High-Altitude Research Vehicle) useless now. We're back to relying on a few nuke-hardened milsats and old-fashioned HF radios. Space Command's still running checks on the Advanced EHF and Wideband Global Satellite birds," referring to the Air Force's newest high-bandwidth communication satellites.

Not good, Aster thought. High-frequency, or HF, radios were 1930s-era long-range communications technology, subject to signal dropouts and considerable weather interference. With the atmosphere charged up from the nuke blast, HF comm would be a noisy fur ball of static. *Better pray that AEHF and Wideband are still with us, or data flow to and from PACOM will be down to snailspeed.* "So, can we get *any* high-speed data to and from those carriers?" Aster asked.

"Not through our 'Battlestar Galactica' comsats, like we normally would, but we *are* using some commercial Direct Broadcast Satellites, sir. That's just one-way comm, though. No uplink back through them. The Air Force and Navy comm troops are trying to get some contingency workarounds in place, but that'll take awhile. Hopefully, the new high-bandwidth sats rode out the nuke effects; just don't know yet. Right now, we're trying to pin down what the hell's working and what isn't," Forester said, a tad testy. Long months of on-the-edge tensions had taken a visible toll on STRATCOM's battle staff.

Aster rubbed his chin absently, struggling to grasp the big picture and range of unfolding potential consequences. He'd better get to the Secretary of Defense immediately. President Boyer would soon be besieged by Paul Vandergrift, the National Security Advisor, who almost certainly would be urging the President to strike back immediately. SecDef T. J. Hurlburt needed ammo to counter any knee-jerk orders from the Oval Office, at least until STRATCOM could sort out this latest nightmare.

That Vandergrift bastard's really going to be screaming for blood now—and the president just might listen to him, Aster realized. He searched the command center quickly. Spotting his aide, Burner

Burns, hovering nearby, Aster crooked a finger in his direction and headed for the door. As usual, the powerfully built Burns was carrying the aluminum-cased "football," keeping STRATCOM'S go-to-war launch codes close at hand. Burns's counterpart in Washington was shadowing President Boyer, as well, carrying an identical case. Hopefully, the president hadn't succumbed to Vandergrift's undoubtedly convincing prattle and cracked the seal on *his* football, yet. . . .

STRATCOM WARGAMING CENTER

"Ladies and gentlemen, we are confirming detonation of a North Korean nuclear device in the extreme upper atmosphere. Initial reports indicate a Taepodong-four launch with warhead activation approximately one hundred thirty miles or so above the surface of the Earth. There are no indications of additional launch activity at this time."

Colonel Jim Androsin, STRATCOM's Wargame Center commander, highlighted the detonation and EMP-impact zone on a large-screen display of Earth and the upper atmosphere. Its image was replicated on small individual displays at every wargamer position in the amphitheater. Using a laser pointer, Androsin pointed at an area of the globe, continuing in grim, clipped terms.

"And this is *not* part of our wargame, folks. Note the yellow band in this area of the Western Pacific, extending over part of mainland China, North and South Korea, Japan, and encompassing the entire Sea of Japan, the Yellow Sea, the East China Sea, and an area east of the island of Honshu. We expect to see the most pronounced EMP effects between the thirty-eighth and thirtieth parallels, or from an east-west line just north of Seoul to here, just south of Shanghai," he explained, encircling an extensive area with the laser beam. "Affected land areas will include all of South Korea, most of North Korea, the Japanese islands of Honshu, Shikoku, and Kyushu, as well as the eastern provinces of China, including the cities of Qingdao, Xuzhou, Nanjing, and Shanghai."

Androsin stepped into the "Pit," a high-tech lectern at the front of the amphitheater, before adding, "Note the anticipated effects BOYD has projected. You can call up those effects on your personal stations, and BOYD will update them with data as status reports come in. Real-world or confirmed information is highlighted in red, as opposed to anticipated effects, which you now see in dark purple. But a word of caution: Real-world info won't be very timely. The EMP has created havoc with comm and data systems in the affected areas."

"BOYD" referred to a powerful, advanced-technology computer system formally known as "BOYDTRIX," which automatically swept up prodigious amounts of information from hundreds of public and government databases, news feeds, and Web sites. It also sifted through that staggering amount of diverse information and computed thousands of possible "connections" per second, distilling them into understandable projections to aid decision making. STRATCOM had pioneered the BOYD system, thanks to a forward-thinking team: Colonel Jim Androsin and his brilliant key analyst, Jill Bock.

The wargaming center was hushed, but not silent. Civilian executives from several communication satellite companies, a NASA troubleshooter, engineers with intimate knowledge of certain spacecraft, officers from all the military services and several allied nations, and representatives from a number of U.S. federal agencies crowded the amphitheater, huddled in small groups. Still registering shock and disbelief, several groups immediately began to address the unfolding calamity. Others hovered around workstations and screen-keyboard combinations sunk into long, curved tables throughout the center.

This eclectic team's latest wargame iteration had evolved to an unprecedented level, no longer virtually "fighting" a war twenty years into the future. Their original "Deadsats II" wargame had been disrupted a year earlier, when a drug cartel–funded Russian maser system had started systematically destroying U.S. and allied satellites. Fed by intelligence from the National Air and Space Intelligence Center and the Joint Space Operations Center (JSpOC), the wargame had quickly morphed into a very effective tool for dealing with a real-time, real-world space war.

Since then, its players had reconvened periodically, examining a complex array of potential "Red," or enemy, moves, and developing options for friendly, or "Blue," forces—the U.S. and its allies—to counter Red and regain an offensive posture. The wargame's participants had been recharged by knowledge that the "alternate futures" and various options they'd developed over the last year had often wound up on the president's desk in Washington. Although the 'gamers were buried inside a bomb-proof underground vault in the middle of Nebraska, they were fighting a very real war, albeit with their collective intellect and judgment—and having immediate, positive impacts.

This latest development, though, had visibly shaken many of the wargamers. A nuclear detonation, even at high altitude and devoid of human casualties, triggered a legacy of horror and fear. To some, it prompted unwarranted guilt, as well. Hiroshima and Nagasaki remained stark reminders for any American who had dealt from the nuclear deck during his or her career, and age wasn't a factor. True, a nuclear explosion had never been more than a theoretical possibility to young officers and staffers now filling the wargaming center. But a few of the older wargame participants could recall atmospheric "atomic" weapon testing by America and the Soviet Union in the 1950s and 1960s. Towering, boiling mushroom clouds from those tests, often featured on television and in newspapers, were unforgettable memories.

A particularly devastating test in July 1962, code-named Starfish Prime, had involved detonating a 1.4-megaton hydrogen bomb at an altitude of 250 miles over Johnston Island in the Pacific Ocean. As a result, electronic systems failed throughout the Hawaiian Islands, some 800 miles away. The Soviets experienced similar widespread failures of overhead communications and power lines, as well as myriad electronic devices, when they conducted their own high-yield upper-atmospheric test.

Now, in 2011, the theoretical suddenly had become all too real. Eyes wide in disbelief, Navy Commander Mark Summers pulled up a chair near Jill Bock, the STRATCOM Wargaming Center's attractive decision-support expert and BOYD analyst. She was seated at a console not far from the amphitheater's Pit. "Hey Jill. You believe this? A *nuke*! I always thought those North-commies might do something dumb, but. . . ."

"Not dumb," she interrupted forcefully. "Desperate, maybe. But not dumb." Bock continued working, fingers flying across a computer keyboard and jabbing a touch screen, her striking green eyes locked onto myriad displays and on-screen windows. Her job was to coax data from a staggering number of sources, all funneled through BOYD, STRATCOM's data-fusion computer system. An icon of her generation's horde of multitaskers, Jill routinely carried on a conversation while herding data through BOYD's voracious processors.

"Well, intentionally zapping themselves qualifies as pure dumb-shitness in my book. Their stuff isn't nuke-hardened like ours," Summers pressed. "Some of our Humvees and the like might lose engine-control chips, but our good stuff, the strike planes and fighters, are all hardened against a nuke blast. I've got a buck that says *they* rode out this EMP hit," he added, a caricature of smug youth and American invincibility.

There was hardly time for debate or lecture, at the moment, but Bock felt compelled to set the good-looking naval aviator straight. "Look," she snapped, facing Summers. "Ignore all the high-handed geopolitical stuff that you'll see flying around here very soon, and pay attention to operational and tactical issues. We're damn near deaf and blind in the Pacific! Most of our space assets are kaput, gonzo, checked out. They're gone, Mark. Comm, nav, weather, direct-broadcast satellites: Color most of 'em cold and dark."

"Yeah, but our *military* stuff can. . . ."

"Mark! Listen up! We're not out of the woods yet!" she barked. "Our military *stuff* was hammered by gamma rays colliding with air molecules, creating Compton electrons. Those Comptons interacted with the Earth's magnetic field, producing an intense electromagnetic pulse. When the EMP and X-rays blasted through satellites—anybody's satellites—they excited and released electrons as they penetrated the guts of those birds. They got into systems like solar-power cells, and they beat up on signal and power cables, attacking the electronics. We've got a high-tech burial ground circling the Earth right now! And that doesn't count all the airborne and ground-based *stuff*, even your precious hardened air-plane 'tronics. Things like engine computers and automatic flight controls can be totally fried, and a hell of a lot of them *are* crispy critters now. *Guaranteed!*"

"Okay, okay! Cut a little slack, lady! How would normal folks know about all that after-the-blast nuke crap? You're talking to a dumb fighter pilot! What you're saying is, we're tits-up in space?" The officer cringed and grimaced, regretting his words the instant they'd cleared his lips.

Bock ignored the pilot's vulgarity and answered calmly, "Not en-tirely. We *have* hardened some of our key military spacecraft, like Milstar, the big DSCS, and new broadband comm birds, and most of our highly classified national technical means platforms. For the most part, those will probably still be okay. 'Course, a few more could degrade or die over the next few weeks, thanks to aging 'tron-ics on the older birds.

"However, there's one giant unknown out there. A lot of our aircraft and ground systems have been updated in recent years with commercial-grade electronics that are *not* rad-hardened," she added, eyes flashing and cheeks flushing with anger. "Dirty little secret there, Commander Summers. But that's what's been happen-ing, even to your precious fighters—at least on some of the older F-18 Hornets—and other *stuff* sitting on your flattops. Saving the almighty buck again! 'What the hey? The Cold War's over! No-body's *ever* going to resort to nukes, so why waste bucks on rad-hardening?' Every Pentagon acquisition wonk's been whispering that crap and getting away with the go-cheap philosophy for twenty years. 'Course, that's precisely what Congress and the administra-tion *wanted* to hear, because they had a massive financial crisis on their hands! They were too busy bailing out Wall Street and car companies to worry about satellites! But we're damn sure gonna pay the bill now!

"Point is," Bock added, switching gears again, "we're in no shape to fight in that area of the world. That's why the North Koreans can't be dismissed as 'dumb.' We have two carrier groups in the northern Pacific, and for some unknown reason, the North's 'Dear Leader' apparently thought they were coming after *him*. If we had been . . . Aw, forget it! Suffice to say that Kim-boy just leveled the playing field by whacking our space *stuff*. Our forces depend on space-based resources of all kinds, and we just lost a boatload of 'em . . . and other *stuff*, as well." She hesitated, both hands motion-less, suspended above her workstation keyboard, then glanced side-ways at Summers and raised one eyebrow. Summers offered a goofy grin, captivated by the beauty of those fiery green eyes.

"Long time ago," she added, "I saw a movie called *Wait Until Dark*, where a blind girl shuts off the lights so bad guys, who're try-ing to kill her, are just as blind as she is. Kinda evened the odds. Guess what. That '*dumb*' North Korean 'Dear Leader' saw the same damned movie."

2 CONFUSION

AIR FORCE SPACE COMMAND HEADQUARTERS

General Erik "Buzz" Sawyer, commander of U.S. Air Force Space Command, ran a hand through short-cropped, salt-and-pepper graying hair and reread the summary slide on a large flat-panel screen mounted on the wall of his office. An Air Force major, who had just completed his briefing, stood to one side of the screen, holding a laser pointer. Tight-jawed, he waited for the four-star's inevitable questions, hoping he'd have the answers. With a room full of bird-colonels and stars, the major's answers easily could make or break the man's career.

"So . . . what you're telling me, basically, is we're screwed," Sawyer growled. He was staring at a one-star general to his right, Space Command's director of operations. The briefing officer breathed a silent prayer of thanks.

"Uhhh . . . I . . . I guess you could say that, sir," the ops director stammered, caught off guard. "As the major said, that nuke blast killed several spacecraft immediately, and the experts say we'll lose at least some functionality on most of the rest over the next month or two. 'The peak impact of a high-altitude nuclear event will occur thirty to sixty days after the event,' according to our Los Alamos

Lab friends," the brigadier general summarized, reading from a document.

"Yeah, yeah. I get it," Sawyer grumbled, still running a hand through thinning hair. He swiveled his chair and eyeballed a crystal-blue sky and the breathtaking scene of snow-covered Pikes Peak, framed by a huge window. One combat boot-encased foot tapped rapidly as he mentally sorted through various options for response. The room's half-dozen or so officers waited silently. Sawyer finally turned, set both elbows on the table, and carefully interlaced rough, knobby fingers.

"Folks, we're in a cascading disaster here," he began, speaking softly. "Losing a handful of birds and the International Space Station to that damned Colombian cartel's two-bit maser last year hurt us, and we've been crippled since then. Y'all know that. But this Korean nuke literally knocked the stew out of our national security space infrastructure. Comm, nav, weather, intel, surveillance—damn near everything in orbit's taken a hit. The charged-up ionosphere's playing havoc with Milstar, GPS, weather, and all sorts of other sat-downlink signals and. . . . Hell, we've beaten this horse enough." He waved a hand in dismissal.

"The question now is: What do we do about it? Obviously, reconstitution of disabled and severely degraded space capabilities has gotta be our first priority. Thanks to severe federal budget problems, we've been forced to get by on workarounds for almost a year. But the camel's back just broke. Every combatant commander in the Department of Defense will be screaming for better overhead support, 'cause God knows where the next dung-storm is going to erupt. All those commanders *have* to be ready to fight, if the president says '*Go!*' North Korea, Iran, and China are probably the most critical areas, so we're gonna focus on taking care of Central Command and Pacific Command first. Let's get creative, people!" Sawyer barked, slapping the tabletop. He scanned his senior staff, eyebrow-hooded eyes holding each man and woman accountable. "This is where we earn our big bucks," he smiled, eliciting a ripple of light laughter.

Two hours later, Sawyer verbally summarized a dozen or so options the staff and he had worked through, arguing the merits and drawbacks of each. Presented on an electronic "white board," the hand-scribed list of items, each accompanied by the name of a responsible action officer, was being printed out for their use.

The four-star chief of all U.S. Air Force space "warfighters" was in dire need of a nature break, thanks to several cups of coffee he'd downed during the lengthy brainstorming session. The bitter taste in his mouth was yet another indicator Sawyer had pegged his caffeine meter for the day.

"All right, let's get at it, folks," he concluded. "Joe, I want to know when we can get those two prototype Naval Research Lab multifunction space platforms launched over the Pacific. We have two Navy carrier groups out there, pretty much running blind now, and several admirals are screaming for overhead coverage. A bunch of sailors and marines are literally sittin' ducks on those boats. They were depending on satellites and that stopgap HARVe near-space platform we parked overhead at sixty thousand feet. Since the nuke blast, they have damned near nothing in the way of comm links and overhead situational awareness. They need help pronto. I want something encouraging to tell those admirals before sundown today."

The colonel assured Sawyer he'd have news from the Naval Research Laboratory soonest. Sawyer concluded the meeting, waited for his staff to depart, then made a beeline for his private lavatory.

WHITE HOUSE SITUATION ROOM

President Pierce Rutledge Boyer, though shaken to his core by North Korea's high-altitude nuclear detonation, fought to portray the image he knew was absolutely essential: a grim-faced, strong, determined leader of the world's only superpower. Americans expected that of its chief executive, especially during critical situations. *Lately, there have been far too many critical situations*, he thought.

At the moment, Boyer's audience comprised five men, all trusted members of his national security team. But soon, he'd have to face the nation's millions of anxious citizens, albeit via TV cameras, explaining why North Korea had "gone nuclear."

"All right, let's get to it," Boyer snapped, his penetrating gaze sweeping the group seated at the Situation Room's large conference table. On a wall-mounted display opposite the president, oversized images of two additional men stared back. Boyer had asked the two—U.S. Air Force General Howard Aster, commander of U.S. Strategic Command, and Admiral Stanton Lee, a retired Navy officer and a key leader of the recently reconstituted Deadsats II wargame underway at STRATCOM headquarters—to join the White House discussion via teleconference.

"We all know North Korea detonated a nuclear weapon at the fringes of space," Boyer clipped, speaking rapidly, "but allow me to summarize the situation. The electromagnetic pulse from that explosion knocked out electrical power grids and electronic systems across a huge swath of the Pacific region. That EMP and other radiation effects also disabled or crippled any number of our national security, weather, navigation, and commercial satellites. We're still waiting for information about the specifics of which systems were affected. However, the collective impact of the losses is still being assessed, right, Howard?"

Aster's image on the big screen nodded. "That's right, Mr. President. General Buzz Sawyer's Space Command team is still tallying up the damage. I might add that Buzz also is in the process of getting some interim-capability satellites into position over the Northern Pacific. They should be in place within the next few days. That'll give us some desperately needed connectivity with two carrier groups we have out there, as well as some limited look-down imagery. But we'll be hurting for lack of space-based capabilities for some time, I'm afraid, sir." He didn't bother stating the obvious: The nation's space infrastructure had been in a severely degraded posture for almost twelve months.

"Understand. Keep T.J. apprised of those new NRL platforms' launch status," Boyer ordered. His hard gaze swept the room, before continuing. "Gentlemen, I want to know what this nuclear disaster really means to the United States, and what action we can and should take. Finally, I want to know the combat readiness of our remaining satellites and those carrier forces in the Pacific as soon as possible." The president glanced toward his Secretary of

Defense, T. J. Hurlburt, who tipped his head in acknowledgment, before replying.

"Mr. President, taking immediate action isn't an option, at least until we reestablish solid communications in that part of the Pacific. We also lack a good operational picture of the situation out there. Anything we do, without having that information, is fraught with significant risk. On the other hand, I can assure you the Pacific Fleet is ready for immediate operations, if necessary," Hurlburt added confidently. "We currently have limited connectivity with the two carrier forces near China. Based on what we *have* heard, all those boats and their airplanes seem to have ridden out the EMP pretty dad-gummed well. We'll know more after those two NRL spacecraft are up and we have high data-rate communications restored."

A retired four-star Army general, who had served as Boyer's defense advisor during the 2008 presidential campaign, Hurlburt had been asked to take the SecDef job, following Boyer's surprising election victory. Had he foreseen the widespread and still-escalating calamities triggered by the takedown of several U.S. satellites and the International Space Station in the spring of 2010, Hurlburt might have politely turned down the president's request.

Right, he mused silently. *Duty calls and old warhorses answer.* As much as he hated dealing with the current nightmares America faced, a soul-deep, lifelong commitment to his nation had never wavered. Hurlburt was better prepared for these challenges than most, and damned few others in Washington could claim the level of credibility and widespread trust he enjoyed. This president at least listened to the retired general, and the Pentagon considered Hurlburt one of its own.

"What in God's name prompted that idiot to pop off a nuke? And why now? If he knew an EMP would knock out most of his own electrical power—and who knows how many computers throughout North Korea—why the hell would Kim sacrifice one of the few nuclear weapons he's been hiding? Surely he knew the world community would retaliate! Somebody kindly explain the logic in that, puh-lease!" Boyer demanded, exaggerating his trademark Southern Virginia accent.

An awkward silence hovered over the room. Most of the men averted their president's hard eyes, some studying folded hands or documents. "Mr. President, if I may . . ."

"Yes, Admiral Lee. By all means." Boyer waved impatiently at the wall screen. Lee had once headed Pacific Command, the largest and most politically complex of all U.S. combatant commands. After retiring from a stellar Navy career, he had served three years as the U.S. ambassador to China under the previous administration. Lee was the quintessential warrior, as well as a skilled diplomat and deep-thinking philosopher. His knowledge and understanding of the Asian mind was legendary.

The elderly statesman-admiral tilted his snow-white cranium, then answered in measured tones. His odd, rather formal, professorial manner of speech prompted a half-grin or two, but also commanded attention. Every eye in the Situation Room was soon focused on the retired officer.

"Consider recent events from two perspectives, that of China and that of North Korea. We are certain that China has kept the North's military leaders apprised of America's satellite losses, as well as the ramifications of those orbital casualties. Further, we know that China, therefore North Korea, are aware of America's heightened military-readiness posture," Lee said. "And Kim surely watches the same CNN and Fox News reports that we do. So he's aware of Iran's embarrassing, failed attempt to attack our air base in Italy; the horrific, televised execution of Canada's Ms. Adkins in Tehran, and the terrorist attacks in Denver and across Europe. Clearly, Iran is front-and-center on the world stage, subjected to withering criticism from several quarters.

"Now, as we all know," Lee continued, somewhat condescendingly, Boyer thought, "China depends on Iran for roughly fifteen percent of its imported oil. In turn, North Korea depends on China for fuel, food, and other essentials. If China becomes concerned about its oil supply, imagine how that concern is magnified in North Korea. We have to assume Kim and his cohorts are quite agitated, given Iran's growing troubles of late."

Lee paused, gathering his thoughts. "Mr. President, for several good reasons, such as Iran's antics, plus intelligence reports about increased military activity in China and North Korea, you recently deployed two carrier groups to strategically vital regions of the Pacific," Lee stated cautiously. Boyer nodded, but didn't respond. "From our standpoint, those carriers are insurance, a measured means to discourage Kim and China's Chairman Yi from exacerbating the situation.

"However, based on a new assessment conducted by my Red team here at STRATCOM, those carriers may be viewed quite differently by China and North Korea. That much U.S. naval and air power at their doorsteps could easily be interpreted as a prelude to surgical strikes, at the least, or a full-fledged invasion, at the worst. North Korea, in particular, could be influenced to respond in a preemptive fashion. I suspect it took very little prodding to convince Kim that his regime was at risk, thanks to those carrier groups. Hence, the Taepodong launch and subsequent detonation at high altitude."

"Admiral, I hear you," Boyer interrupted. "But I find it very hard to believe that China would condone Kim torching off a nuclear warhead a hundred thirty-seven miles above his *own* country, knowing full well China also would be nailed by the EMP! Especially when China supposedly concurred back in oh-seven that the North had abandoned its nuclear program. Surely, Chairman Yi and his people would have blocked any Korean attempt to employ nuclear weapons this way! That's simply too over-the-top provocative at a time like this, particularly with the global economy in shambles!"

Lee lifted his chin, half-nodded, and continued. "Agreed, Mr. President. But I suspect China had no forewarning of the North's plans. Yes, the Chinese may have monitored preparations of the Taepodong missile, but probably did not discourage the so-called 'test launch' that Kim had announced. Although provocative, a long-range missile test—if successful this time, of course—would have served China's interests quite well. It would divert U.S. and United Nations attention away from Iran—which China quietly supports—and underscore the perception that America is wounded, in a space sense, and cannot monitor North Korea adequately.

"But there's another important element here, sir," Lee added, leaning closer to the camera. "We believe China also was alarmed by our deployment of those carriers. For years, People's Liberation

Army generals have been warning that America could choke off the flow of China's steady string of oil shipments through the Malacca Strait. Our Navy has patrolled those waters between Malaysia and Indonesia for years, ostensibly to fend off pirates and ensure Mideast oil reaches Japan and other U.S. allies. However, eighty percent of China's imported oil passes through the Strait, and as the Chinese economy boomed, those oil shipments became incredibly critical to the nation's national and economic security. The PLA's number-one question of late has been, 'What would happen if the U.S. closed the Strait, as a means of discouraging a Chinese attack on Taiwan?' The answer is clear: The impacts on China would be immediate and profound. One PLA general even claimed it would trigger war with America."

"Excuse *me*, Admiral! Are you insinuating that the president's order to move those two carrier groups is responsible for North Korea's absolutely insane launch of a Taepodong, an action that has brought the entire world to the threshold of nuclear war?" huffed Paul Vandergrift, Boyer's National Security Advisor and director of the National Security Council. "That's approaching insubordination!"

"Paul, for God's sake, can the gratuitous suck-up!" Boyer retorted. "I *have* to understand what the hell is going on here, and the admiral's expertise on Asian matters exceeds yours considerably! Admiral Lee, my apologies for Mr. Vandergrift's indiscretion. Please continue."

Vandergrift's feigned indignity had been transparent to even Boyer, who usually tolerated such tail-kissing from his National Security Advisor. But the latter's disastrous, almost-adolescent, off-the-books maneuvers last year—particularly the foiled "cyber-worm" attack against Iran's air defense system, followed closely by using the late Charlotte Adkins, a Canadian official, as a backdoor emissary to Tehran—had severely maimed Vandergrift's credibility in the White House. Why the arrogant jerk still occupied an office there was beyond T. J. Hurlburt's understanding.

The idiot's gotta have incriminating pictures of Boyer, the SecDef groused silently, disgust evident in the dark scowl he shot Vandergrift. The two men had almost come to blows after Charlotte Adkins's televised firing-squad execution at the hands of some Iranian hothead the CIA had code-named "Dagger."

"Thank you, sir," Lee acknowledged. "My intent is to convey a rationale for the North Korean's nuclear ploy, not assign blame. Now, if I may return to China . . . When two of our Pacific Fleet carrier groups started moving, China's hawks almost certainly sounded the alarm in Beijing. Chairman Yi was probably inundated by PLA warnings: 'Just as we feared, the Americans are preparing to close the Malacca Strait, assuming an offensive posture near our homeland!' they surely cried. With very few overhead intelligence satellites of its own, China had limited means to determine our intentions. I suspect a measure of that Chinese uncertainty was conveyed to Pyongyang, as well."

"Do you think Yi or the PLA encouraged Kim to launch that nuke?" Boyer asked.

"Sir, I do not," Lee responded firmly. "I've not seen a shred of intelligence that indicates China had any inkling North Korea might resort to a nuclear ploy. Further, if they had seen any hint of such a drastic act in the making, I believe the Chinese leadership would have gone to great lengths to ensure the North's launch was scrubbed. I have seen absolutely *no* signs of collaboration between China and the North Koreans concerning this incident."

Boyer turned to his National Intelligence Director, Herb Stollach. "Herb, any comments? Do the admiral's interpretations jibe with your info?"

"I concur with Admiral Lee's assessment, Mr. President. Our agencies closely monitor all communications traffic between China and North Korea, and none of them picked up even a whiff that would indicate China knew what the Koreans were up to. In contrast, traffic from China *since* that nuke went off has been hot and heavy. Chairman Yi personally blasted Kim in not-so-diplomatic terms. The National Security Agency code breakers are working overtime, but what we've seen, so far, tells me Chairman Yi and his people were caught completely off guard. Hard as it is to believe, the Chinese didn't see this coming—or at least Yi didn't."

"Cripes! So Kim and his mavericks dreamed this up on their own? Because they thought *we* were going to invade the North?

Why? To steal their last bag of rice, for Heaven's sake?" Boyer asked, incredulously, throwing both hands in the air. "Look, I don't pretend to understand how that unpredictable nutcase thinks, but Kim is anything but stupid. Something spooked him into going for broke, forcing him to expend one of his precious few nukes this way. And two U.S. Navy carrier groups cruising miles away from Korean shores aren't enough to push him over the edge.

"I'm sorry, gents, but this old Virginia lawyer does *not* see a solid case for the North going nuclear. And until we can determine what the hell led to Kim's unprecedented, practically suicidal decision to explode a nuclear weapon over his own country, I'm very reluctant to respond, politically or militarily. Gents, I need answers!

"One more thing," Boyer added quickly, redirecting the discussion. "Is intel picking up indications that Iran might be capitalizing on our predicament in the Pacific? I'd expect Ahmadinejad and his snakes to do just that, maybe in Iraq or Afghanistan. Herb, what are your spooks picking up?"

"A lot of chatter, but nothing definitive," Stollach answered. "We're hurting for overhead assets, of course, so our imagery and signal intelligence collection is intermittent and spotty, at best. Consequently, we're dealing with a thousand clues, you might say, but we have yet to solve the mystery. It's hard to build a big picture from those sporadic intel clues. But the 'Company' has a deep-cover, in-country humint asset that seems to be onto something big. We're trying to get a handle on whatever that . . ."

"By 'humint,' you mean that agent who said Iran has acquired a Russian missile-guidance package?" Boyer interrupted.

"Right. A Glonass guidance system," Stollach affirmed. "That's the Russian version of our GPS navigation-and-timing satellite system. We know a sophisticated Glonass receiver-guidance system was delivered to an Iranian agent codenamed 'Dagger.' The CIA believes he's the same devil that had Charlotte Adkins executed in Tehran. We know the guy is in tight with Iran's power base, both the radical-Islamist mullahs and President Ahmadinejad. And he's a mean SOB. If Dagger has a Glonass guidance system, you can bet it'll wind up on another Shahab missile. And it won't miss this time," Stollach concluded ominously.

Boyer grimaced, then turned to the big screen. "Howard, I trust your wargamers are exploring the ramifications of Iran launching another missile. What's their best guess about potential targets this time?"

"Yes, sir, they've taken a hard look at that. As of a few hours ago, their consensus was not pretty, either," Aster said, lips drawn to a thin line. "We have reason to believe Iran is preparing to launch another Shahab-4, probably armed with a nuclear or biological weapon. They can't have many nukes left, so . . ."

"Because you think Speed Griffin's 'Rods-from-God' mission wiped out Iran's nuclear-development site at Parchin," Boyer interjected as a statement, not a question. Months before, Brigadier General Hank "Speed" Griffin, a test pilot currently assigned as the U.S. Air Force liaison to the White House's National Security Council, had flown the supersecret two-stage "Blackstar" system's small XOV spaceplane into low-Earth orbit and fired two hypervelocity missiles at Iran's Parchin nuclear facility. Basically rocket-powered titanium rods, the missiles achieved hypersonic reentry speeds, which translated to massive releases of destructive energy, when they slammed into the Earth.

Even though these "Rods-from-God," as the Blackstar test team had dubbed the missiles, carried no warheads, their high-Mach-number velocities ensured a level of destruction comparable to that of a small, tactical nuclear weapon. Tremendous amounts of kinetic energy, expended in a very brief span of time, had reduced Iran's prized underground nuclear research and development site to smoking rubble, yet left no residue that could be traced to the U.S.

"Affirmative, sir," Aster said, nodding. "Prior to Speed's mission, that CIA agent in Iran had reported seeing a warhead being loaded onto a Shahab just before Iran's rocket forces lobbed that nuke at Italy. Since then, our intel folks have monitored the same launch area in the mountains—thanks to the few space-based recon assets we still have—where mobile Shahabs almost certainly are hidden in tunnels. And they've seen a flurry of activity at one site. In short, sir, the intel community and our 'gamers think Iran may very well be prepping another missile armed with a nuke or bio weapon.

"As to where it might be going, there's a broad range of possibilities," the general continued. "But our guys don't think Ahmadinejad or his number-one boy, Dagger, will waste one of their few nukes on an American Air Force base. They'll go for the jugular this time."

Boyer waited, but Aster only stared from the big screen, his jaw muscles twitching. "And your team's best guess at what that jugular might be, General?" Boyer said, with an edge of irritation.

T. J. Hurlburt read Aster's hesitance and stepped in. "Mr. President, that assessment should be made by Herb's intel team, not Howard's wargamers."

Boyer shot his Secretary of Defense a withering glare. "Hell, I don't care who *should* be making assessments! I need every damned 'maybe' and 'what-if' on the table! Let's have it, T.J.!"

Hurlburt looked to Stollach, then back at Boyer. "Either a large European city or Israel, sir. Take your pick."

Boyer grimaced and faced Stollach. "That as good as it gets, Herb? Surely your people can be more explicit."

" 'Fraid not. We've lost so many satellites that we can't monitor Iran as well as we used to. And, as you know, our on-the-ground intel resources are minimal, at best. We see activity in the mountains, roughly where that last Shahab was launched, but we have very few comm intercepts to work with. Nobody's talking on the radio or wireless phones. Something big's stirring out there, sir, but we don't know exactly what they're doing or why," Stollach said. "My assessment? Iran's going to strike, and soon. Where and when, I can't tell you at this moment. But I suspect those Persian wild cards intend to make a considerable splash this time."

"Damn, damn, *damn!*" the president half-whispered, as he turned back to Aster. "Any more cheery news, General?"

"No, sir. As new intel comes in, we're cranking it into the wargame immediately. I'm personally relaying our updated insights to the Secretary, as soon as they're vetted here."

Boyer stood, signaling the meeting had ended. "Many thanks, General. And I appreciate *your* insights, too, Admiral Lee. Your wargamers out there in Omaha have proven to be invaluable throughout this nightmare. I'm still depending on you. Keep at it."

He waved and headed for the door, his chief of staff and Paul Vandergrift in tow.

Hurlburt glanced at Aster's and Lee's grim-faced images, shot them a thumbs-up, then turned to follow Boyer.

Aster waited until the video link blanked, then turned to Lee. "Admiral, I have to say, you're one gutsy guy. Challenging the president on his decision to deploy those two carrier groups . . . ," Aster said, shaking his head admiringly.

Lee stared at his folded hands a long moment, then spoke softly. "Howard, at this moment, I feel our nation is in grave danger. We cannot afford to misread our adversaries' motivations, let alone take action that might lead to irreversible, unintended consequences. Whatever our president does next, it must be absolutely correct. There's no room for error or second chances. We owe him the very best, unvarnished advice and information we can provide."

Aster nodded in agreement. "Roger that. Let's hope those closest to him see it the same way. By the way, I advised *against* deploying those carriers. Paul Vandergrift talked the president into it. That damned Vandergrift is dangerous!"

"I know," Lee sighed, staring at the static-filled screen. Suddenly concerned about how weary the stately retired naval officer looked, Aster stood and gently grasped Lee's shoulder.

NATIONAL RECONNAISSANCE OFFICE

Hours of meticulous, detailed work had removed the last vestiges of doubt. Imagery-intelligence analysts—professionally trained skeptics—had compared data from two sources, an aging Lacrosse radar satellite and Hawkeye, a still-secret developmental, hyperspectral-imaging spacecraft flying in an oblong, highly elliptical orbit, were in rare agreement. The analysts had concluded that Iran was, indeed, preparing another Shahab for launch. And if the National Security Agency was interpreting sketchy communications intercepts correctly, Iran was embarking on a very bold and potentially catastrophic campaign. Taken together, the imagery and comm data built a compelling, scary picture.

Communications intercepted by NSA listening posts had snagged a discussion between a couple of high-ranking officials in Iran and North Korea. One gravel-voiced Iranian had boasted that U.S. space capabilities were crippled to the point of irrelevance, assuring a North Korean general that "the infidels are now blind." A maser fired by a Russian scientist from Tajikistan had "destroyed Satan's spy satellites and space station," he claimed.

Erroneously, he had trusted Iran's crude encryption equipment and protocols, believing they precluded NRO signal intelligence satellites and NSA computers from gathering, then decoding, his scrambled phone conversations. Consequently, the Iranian had goaded his North Korean security counterpart to "finish gouging out the imperialists' eyes, before your country is invaded from the sea. My people will finish the task."

The Korean general, obviously ultra-concerned that such an invasion was imminent, assured the Iranian that "Korea will strike first!" Evidently, the two countries intended to capitalize on the West's wounded space infrastructure. Shortly thereafter, North Korea had launched its nuclear-armed Taepodong missile.

Finally, the NRO's most-senior analyst summarized the findings of both agencies, and concluded the NRO-NSA teleconference. He then dialed the NRO's deputy director. A woman's voice answered.

"Ma'am, we've confirmed the Lacrosse radar pickup, and we've cross-checked it with Hawkeye imaging. Our conclusions dovetail with NSA's, based on a few 'golden' intercepts they caught prior to the North Korean Taepodong launch. Doesn't look good, I'm afraid. I just sent you a summary, and I'd like to talk you through it, if you have a moment. Look for an e-mail marked 'Shahab-4.' " He waited until she confirmed both the imagery and NSA executive summary were on-screen.

"You've run multi-pass change analyses on these radar and multi-spectral images?" she asked, comparing several before-and-after photos. Heavy-duty computer analyses highlighted significant changes in selected objects, areas, and terrain patterns over a period of time.

"Yes. Absolutely no question, ma'am. They're preparing a Shahab-4 for launch," the analyst answered firmly, then directed his boss's attention to a specific image. "Take a look at the radar time-lapse from last night." Again, he waited.

"Oh, shit," the deputy breathed. Trucks had definitely moved between a suspected weapons-storage site and a tunnel complex identified as missile "barns" by COBI, a CIA deep-cover agent in Iran. "Any confirmation on the trucks' cargo?" She knew the answer, but would need iron-clad proof before taking these findings higher.

"Yes, ma'am. Lacrosse spotted a single warhead, unloaded outside the missile tunnel last night. Several soldiers handled it with the utmost care. And the last image in that Hawkeye infrared series pretty much tells the story."

The deputy groaned. A high-resolution infrared photo, taken from a low angle, showed soldiers working on a Shahab missile's nose. She moved and clicked a mouse several times, scanning through the series of infrared and radar images a second time. *That SOB's gotta be a nuclear warhead!* she concluded, shivering slightly.

"Good work, Sam," she said, trying to sound more resolute than she felt. "Keep everything we've got over the region aimed at that site, and let me know immediately if they start fueling the Shahab."

Sam swallowed hard. "Ma'am, we think they've already started fueling. We'll know for sure when we get the next-pass Hawkeye dump. We're hurting for timely sat-revisits, as you know." Maser attacks the previous year—high-power microwave beams fired by mercenary Russian scientists funded by a Colombian drug cartel—had disabled several military, intelligence, GPS, and commercial communication satellites. Although a quick-reaction near-space platform positioned over the Mideast was providing stopgap GPS service, critical intelligence, surveillance, and reconnaissance (ISR) resources in the region were severely curtailed.

A long silence ensued while the deputy director absorbed the implications of that statement. The long-range missile could be ready to launch within a few days. "Okay. Stay on it. If that damned Shahab rolls out of its tunnel, scream immediately," she clipped, then broke the connection. Her hand trembled slightly as it set up a conference call she once thought would never be made on her post-Cold War watch.

WHITE HOUSE OVAL OFFICE

"Mr. President, I acknowledge your concerns, but you must understand. Israel *will* protect itself, regardless of so-called 'world opinion.' I cannot—*will* not—trust my people's safety and its future to the outside possibility that you or the United Nations or our European friends *might* be able to dissuade Iran from attacking my country!" stressed Moishe Baron. The Israeli prime minister spoke calmly, but there was no mistaking the strength or resolution of his tone.

President Pierce Boyer rubbed a forefinger along his left temple, absently staring at the Oval Office's desktop. He adjusted the phone receiver pressed against his ear and sighed.

"Moishe, believe me. I'd say precisely the same thing, if I were in your position," Boyer said. The epitome of sincerity, observed Gil Vega, the president's chief of staff. Vega and Boyer were alone in the storied office, a subtle measure of the president's growing dissatisfaction with Paul Vandergrift. Conducting a critical discussion with a nation's head of state about a potential attack against that ally, but without the National Security Advisor in attendance, was unprecedented. *The president believes he's on his own!* Vega flashed, suddenly worried. *Winging it solo is how leaders make irrevocable, history-defining mistakes. . . .*

"But think about this," Boyer continued, almost pleadingly. "Elevating your state of military readiness at this time almost certainly will be misinterpreted by Ahmadinejad and his radical generals—to say nothing of the mullahs, for God's sake! 'Israel is poised to launch a preemptive strike!' they're probably shouting. You're giving Ahmadinejad precisely what he wants: a defensible justification to attack Israel, in the name of defending Iran. Regardless of what that madman does, it'll play well in the world press, because Israel will be painted as the aggressor! Don't fall into his trap, Moishe!"

Vega could only hear one side of the conversation, but having met the prime minister on several occasions, he knew how black-and-white the former Israeli defense chief viewed his world. No shades of gray. No room for give-and-take or compromise. Survival of Israel was all that mattered. Opinions and alternate views be damned.

There was a long silence on the other end. "You're undoubtedly correct, Pierce," the prime minister finally said, dropping into the same first-name informality Boyer had adopted. Baron sighed audibly. "But I have little choice in the matter. If an Iranian atomic weapon detonates in Israel, our nation might cease to exist, or never recover. Permit me to remind you: We only have seven million people, and our entire country's landmass is smaller than most of your states. Israel simply cannot 'soak up' a devastating nuclear attack. No; I will not stand by, defenseless, while political maneuvers by those *not* in Iran's crosshairs attempt to discourage madmen from treading their unholy path to eliminate the Jewish state . . . forever."

Boyer tried a different tack. "I understand! I really do. But is there *any* solid evidence that Iran is about to strike? What gives you cause to seriously consider being *so* provocative, poised at hair-trigger readiness? Especially now, when Iran's especially unpredictable?"

Baron laughed harshly. "You know perfectly well I have such evidence! And so do you!"

"I don't follow. . . ." Boyer said, feigning ignorance. Vega cringed, shaking his head enough to catch the president's attention. Boyer raised a hand, palm up, in mock question, and shrugged. Vega looked down and shook his head again. *You're screwed, Pierce!* he thought, but couldn't say, fearing the Israeli prime minister would hear him.

"President Boyer, please, please," the prime minister chuckled condescendingly. "Spare me. Your CIA surely has informed you that the Spanish intelligence service passed along incontrovertible evidence that Iran is preparing a Shahab-4 missile for launch. Your own NRO people have concluded that the missile is very likely armed with a nuclear warhead. With few exceptions, the consensus of your and others' intelligence services is that this missile and its deadly cargo are destined for Israel."

Boyer groaned audibly, again rubbing his temple vigorously. "Good God, Moishe! Is your Mossad embedded so deeply in Washington that you know more about my own intelligence than I do?"

The question evoked only silence. "Damn! Mossad is good, I admit. What else should I know, since you apparently have more inside information than the President of the United States does these days?" Boyer was peeved, solidly checkmated—and obviously embarrassed—by another head of state.

Vega again shook his head, drawing a dark scowl from Boyer. *Damn you, Pierce!* Vega fumed silently. *You left yourself open to that, and got slam-dunked!* Every president was briefed on a simple fact of political life: Intel services the world over shared information. If ever there were a cross-border cartel devoid of 100-percent loyalty to a given nation state, it was the free-world's intelligence agencies. They all shared off-the-record, backdoor information, each convinced that sharing was the only way to keep ambitious politicians in check. Veteran "spooks" could point to any number of Cold War-era instances that, at least in their view, proved intelligence-sharing had avoided nuclear Armageddon.

Vega's gut churned. This telecon was not going well. He drew fingertips across his throat, signaling that the president should end the conversation. Boyer pointedly ignored the advice.

"For one, I suspect that your people will not be able to monitor Iran and its Shahab adequately for much longer," Baron continued, almost gleefully, Boyer perceived, irritated anew. "And a study done about ten years ago by your Defense Threat Reduction Agency predicted that a single low-yield, high-altitude nuclear explosion—a blast about the size of the Hiroshima event—would disable ninety percent of all satellites in low-Earth orbit within a month. Yes, America's highly elliptical-orbit spy satellites might survive, because they're supposedly hardened to resist radiation, but my people doubt they'll function as long as projected. Radiation effects from that North Korean weapon will certainly degrade them, at best, and probably render the spacecraft unusable within thirty days."

"Good Lord! Tell me, please. Just how does an *Israeli* prime minister know so much about the projected life spans of *American* satellites in a radiation environment?" Boyer asked, reluctantly impressed.

Baron chuckled. "I'm no physicist, but I *was* blessed with a good memory. Truth be known, my staff briefed me about the probable impacts of that North Korean detonation on our own satellites. I'm sorry to report that it appears *our* LEO satellites are doomed, as well."

"Meaning Israel *also* will lose its ability to track what Iran's up to, at least from space?"

There was a long pause before the Israeli answered. "That's true, yes. But we have other means. . . . ," he said, leaving the obvious hanging, unspoken. Numerous Mossad intel agents assuredly were buried among Iran's populace, giving Israel much better human intelligence than the U.S. enjoyed.

The 1990s had not been kind to the American humint apparatus, which was largely disassembled in the wake of the Cold War. Political pressures and an administration that had other priorities—such as an ambitious social agenda—had diverted funds once devoted to intelligence-gathering. Those forces, plus an almost-arrogant Pentagon trust in "overhead technical means," or satellite-acquired intel, conspired to quickly dismantle U.S. spy networks around the globe. Rebuilding them after the 9/11 attacks on New York and Washington had been a slow, painful and expensive process, especially within Muslim societies that hated the United States.

"Okay, Moishe. You win," the president conceded. "Yes, my advisors believe we *will* lose a huge percentage of our intel satellites within the next few weeks. Yes, Spain's very capable intelligence service *did* notify the CIA about that Shahab, which confirmed our own 'other' intelligence. Yes, our guys *do* think the missile's been armed with a nuke. But we have no reason to believe Iran will launch it against Israel. Unless you, Mr. Prime Minister, give Iran's radicals a reason to do so. I beg you: Please, for the sake of the entire world—not only Jews, but for the sake of people everywhere—*please* stand down your forces. Do *not* give Ahmadinejad a reason to launch that missile!"

Baron laughed again, a ragged, harsh sound. "Oh, Mr. President! It's so easy to give advice about preserving peace and to speak on behalf of 'the world's people,' as you sit in your comfortable Washington office, thousands of miles from Iran. Americans are not living under the threat of absolute, total destruction, mere minutes of Shahab flight time from Iran! Only Jews are in the crosshairs of that devil, Ahmadinejad, who has sworn openly to wipe Israel off the map!" He hesitated briefly, giving Boyer an opening.

"My God, Moishe! Think! Your actions could trigger exactly what you hope to prevent . . ." Boyer's plea was rudely interrupted.

"Pierce! Do you really think I have not carefully considered all possible ramifications of every potential action and response? *I have*! And I give you my word: Israel will not attack first. But, if Iran launches that Shahab, I *will* respond with the full might of my nation's capabilities, regardless of the Iranian missile's target this time. I simply *cannot* risk the safety of my people! Do *you* understand?" The voice was glove-soft, but the message was hardened steel.

Boyer threw a hand in the air, exasperated. "I understand, Moishe. I do! And I sincerely appreciate your assurance that Israel will not strike preemptively. But I *also* cannot condone a military response to an Iranian launch, should Iran strike. Surely we have grown beyond the barbarism of always slugging it out, killing thousands of innocent people to solve disagreements among neighbors! Civilized people *can* do better, Mr. Prime Minister!"

Again, Baron laughed quietly, before responding as if speaking to a child. "Ah, yes, Mr. President. Civilized people *should* be acting much differently, I agree. However, what do high-minded, civilized people do, when facing a barbaric, *un*civilized enemy committed to ending your very existence? You—*you*, personally—have never been in combat, never come face-to-face with men intent on killing *you*. I have. My father has; my brother has; my entire nation has," he explained, voice catching slightly. Baron sighed deeply again, adding, "We must deal with the world as it is. You, as the leader of a superpower, *cannot* afford to continue seeing the world as you would *like* it to be, instead of as it *is*. May God give you eyes that *see* the serpents and evil that actually are at our door—and at *your* door—as well as the wisdom to deal with them appropriately."

Boyer signed off, slowly cradled the small portable handset, and slumped in his chair. He glanced at Vega, his eyes the definition of despair and stress-induced fatigue. "Gil, this does *not* look good.

Not good at all." Not waiting for a response, the president spun his chair and stared at an immaculate White House lawn outside the Oval Office window. He'd never felt so alone in his entire life. Nor as helpless to prevent catastrophe.

TEHRAN, IRAN

Hassan Rafjani was a very happy man. His cryptic phone conversation with the North Korean general about the launch and detonation of a nuclear weapon had paid off. He'd just received confirmation, via an equally cryptic e-mail message, from a trusted Iranian emissary in Pyongyang: "A new Sun has risen in the east."

Sent by a Revolutionary Guard agent he, Rafjani, had personally recruited and trained, that simple one-liner spoke volumes. An expensive, years-long, Rafjani-conceived campaign had come to fantastically successful fruition. He'd dispatched the agent to North Korea as a personal link to that nation's leader. Whatever the quirky Kim Jong Il wanted, Rafjani made sure the "Dear Leader" received it. Consequently, Kim and his minions ultimately succumbed to the continuous stream of not-so-subtle warnings Rafjani had conveyed. Convinced an attack against North Korea was imminent, Kim had launched the nuclear-armed Taepodong.

Rafjani printed the agent's message and left his spartan office. Walking rapidly down a long hallway adorned with items recalling Persia's storied past, he slipped a hand under a long coat and adjusted a thin, sheathed knife tucked into a belt at the small of his back. He took perverse pride in the "Dagger" moniker the Great Satan's Central Intelligence Agency had given him. It stemmed from a sinister, ripple-edged blade used to dispatch more than one poor soul who dared challenge him. The razor-sharp knife also had loosened the tongue of a suspected CIA agent who had infiltrated a close-knit band of believers that had held American infidels for 444 days, following the Shah's ouster in 1979. At the time, Rafjani had been a young, passionate soldier, who had sworn allegiance to the Holy Khomeini. In the thirty-plus years since, Dagger had become a powerful figure, the brains behind Iran's more-visible leaders. Few people outside the government knew anything about him, let alone the degree of power he wielded.

At the president's office, he waited a few brief minutes before being ushered inside. A smiling President Mahmoud Ahmadinejad met Rafjani with outstretched arms and cheek-brushes, greeting him as a brother. Indeed, the two men had become close allies over the many years, since they were students at Tehran University. The president was dressed casually in tan slacks and a long-sleeved, pastel cotton shirt.

Seated, Rafjani quickly summarized the messages from North Korea, gleefully detailing how his campaign of fear had convinced the hermit nation's political and military leadership that the two American aircraft carriers were planning to invade the North at any moment. The two men laughed at the brilliance of Iran's clever ruse, amazed at how relatively easily Rafjani had been able to fuel Kim's paranoia, while keeping China's army and Chairman Yi in the dark. In short, the North's high-altitude detonation of its sole remaining nuclear weapon was a testament to Rafjani's skill and growing international influence, albeit behind the scenes. Operating in the shadows, yet in concert with the president's public actions, was proving to be extremely effective.

Ahmadinejad's effusive congratulations and praise were sweet music to Rafjani. He had dedicated his life to eradicating the infidel filth that brought shame on him and Persia. The time had come to complete that holy mission.

"My brother, you have done much to hasten the return of our holy Mahdi," Ahmadinejad finally said, leaning back in a massive chair, fingers of his delicate hands joined in a steeple. Rafjani was struck, yet again, by how physically diminutive the man was. "The state of our troubled nation is such that we must take action to hasten the day of final judgment, as was prophesied. Are your warriors prepared, ready to act?"

"Yes. As we speak, our Hezbollah brothers are poised to unleash havoc on America. The first blow will be struck in California," Rafjani explained. Ahmadinejad nodded knowingly, offering his trademark squint-eyed, scraggly-beard smile. He and Dagger had frequently discussed how depraved the so-called "Golden State" had become. Allah would be very pleased with Rafjani's bold plan for cleansing, the president suggested, again prompting Rafjani's humble nod. The two quickly reviewed detailed plans Rafjani and his small band of dedicated foot soldiers had prepared so carefully, then set in motion years ago.

"My President, the California operation will paralyze the weak Americans and turn their attention inward," Rafjani boasted. "Their politicians will quarrel and blame each other for the disaster. They will be licking their own wounds, and fail to see or hear 'Allah's Sword' singing, until it is too late to save the Jews."

Ahmadinejad smiled, revealing nothing. "And you are absolutely certain the American president will not come to the Jews' aid?"

"Ahch!" Rafjani exclaimed harshly. "Boyer and his swine in Congress are the epitome of America's weakness. They believe only in talk, talk, talk, cowering under their beds, hoping and praying the holy power of Islam will simply go away! The Satan will *not* fight! America's moral decay runs deep, the product of stolen Muslim oil and a thirst for pleasure. The resolve that once was America is now nothing but rotting stone and rusting steel."

He stood and stretched both arms wide, his face flushed as he expounded, "Recall how the cowards retreated from Iraq, only too happy to accept Persia's 'help' in Iraq's partitioning! Now, oil flows freely from the west and north, the 'Iraq State of Islam,' feeding the massive appetites of our Chinese friends, while filling our treasury. And the Russian whores grovel at our door, begging to build new refineries for us! Yes, America's days as a superpower are past. *Insha' Allah!*"

He paused, staring down at Iran's president, who was smiling, silently nodding in agreement. "The time to activate Allah's Sword has come," Rafjani said softly, then waited.

Ahmadinejad said nothing. He opened a side drawer of his expansive desk and removed a document. A garish golden seal and crescent marked it as a presidential decree. With a sweeping flourish, the president signed the paper and spun it toward Rafjani.

"I am honored," the latter said, scanning the text. "Will there be trouble with Mohamed?"

Ahmadinejad shrugged. "It is of no concern." If the current head of Iran's al-Quds, the terrorist-training arm of the Revolutionary Guard Corps, objected to Rafjani taking over, Mohamed would simply disappear. Dagger would now be directing terrorist operations abroad, as well as all intelligence-gathering performed by the al-Quds in western nations. Under his leadership, Allah's Sword would draw blood on a scale the world had never seen before.

3 CHINA WORRIES

STRATCOM WARGAMING CENTER

The initial shock that followed North Korea's brazen nuclear detonation soon turned to anger, and anger led to short-lived chaos within the wargaming center. Colonel Jim Androsin had his hands full, trying to get his 'gamers back in the groove. He realized their anger would soon lead to determination, a fierce determination that must be harnessed and channeled properly, or precious time would be wasted.

Admiral Stanton Lee watched the Army colonel nudge and cajol his players. As had been his way throughout an impressive military career, Lee again sensed that he, as one of the wargame's key leaders, must rise above emotions of the moment and keep his mind on the broader picture.

Emotion can lead to distraction, he reflected. The consequences of North Korea's nuke must remain the wargamers' focus. Not only on the blast's material effects, but on assessments of what America's adversaries would likely be thinking. *How will Iran react?* he wondered. *And of utmost importance, how do we sort through the competing forces within China?*

He reverted to a step-by-step mental process to crystallize the current situation, a private mental exercise that began with: *What do we know?* Two carrier battle groups were steaming toward the region of interest, near the Korean peninsula and the South China Sea. Air Force and Navy strategic commanders had increased alert levels across the globe, particularly in the Pacific. Southeastern China had also been blinded by North Korea's irrational high-altitude detonation, and the resulting loss of situational awareness was almost certainly being leveraged by worrisome elements inside China.

Those factions are very dangerous players, who undoubtedly will take advantage of this unfolding high stakes drama, Lee concluded. He shook his head and sighed deeply, a futile attempt to shake off a gnawing fatigue.

He searched the center, looking for the STRATCOM commander. Aster was huddled with a group of officers, reviewing the latest BOYDTRIX assessments flowing in from other U.S. combatant commanders, or "CoComs." As Lee approached, Aster glanced at the ramrod-straight retired naval officer. The Air Force general's eyes were red-ringed and deep lines furrowed the space between his eyebrows.

"Admiral, that nuke damn near slammed us back to the Stone Age! How the hell can we keep track of North Korea, China, Iran, and God-knows-who-else, with our space assets in the shitter?" Aster grumped.

Lee was taken aback by Aster's outburst, airing frustration in front of subordinates. It wasn't like the normally cool, level-headed STRATCOM chief. "We'll get through, General. We always do," Lee said calmly. "Of immediate concern: I believe the President should order an elevation in ThreatCon level for *all* our forces, not just those in the Pacific. Considerable paranoia is sweeping across the globe, and those wishing us harm are poised to recognize opportunities the North has afforded them, and will be quick to leap into the breach."

Lee touched Aster's arm and flipped a subtle side-nod, indicating a private discussion was in order. Aster stepped away from the group huddled around a BOYD 20-inch monitor. "Howard, I am very concerned that certain factions in the Chinese army could be the culprits behind that North Korean caper," Lee said in a low voice.

"But the intel community claims Iran is . . ." Aster began.

Lee interrupted, raising a palm. "I know. I may be wrong, but I don't think so. Badgering the North into launching that nuke would require a much closer relationship than we've detected with Iran. The Iranians provide money, yes, but I don't see North Korea playing proxy for Iran. No, the 'Dear Leader' has something much bigger at stake, and only China's PLA can deliver."

"Admiral, you telling me the Chinese would deliberately blind themselves, turn out the lights in major cities and knock out God knows how many of their computers?"

"No, I'm not. What I *am* saying is there are elements within the PLA willing to put up with a degraded infrastructure to accomplish a larger objective. They'd do it to entice us into stumbling around in their backyard, creating incidents, albeit unintentional on our part. Even triggering strategic responses."

Aster raised an eyebrow. "You mean, *China* going nuclear?"

"Maybe. Hard to tell what the PLA has in mind. But events are cascading rapidly, and we need to be viewing them from China's perspective. To that end, I want to start a side-wargame focusing on China."

"I concur, Admiral. China warrants careful watching. Tell Androsin to make it happen. And keep the damned Russians in your sights, too. They tweak our nose every chance they get! BOYD's scarfing up tidbits about a Russia-Venezuela satellite being prepped for launch. That one has us all wondering. *Ahhh . . . !*" Aster flared angrily, rubbing his face with both hands.

He was bone tired and needed a shower and shave. He could no longer think clearly, and that's precisely when he and others were prone to making mistakes. Big mistakes. Long hours were taking a toll on his wargaming teams and battle-staff troops, as well as their leaders.

"Okay, we'll keep an eye on China," Lee said. "But to your point about the Russians. I'm sure they, too, are waiting patiently in the wings, probing for weak spots. For them, the risks of miscalculation are great, but the potential spoils are too lucrative to ignore, especially if we get bogged down with China. However, unless the survival of Mother Russia is at stake, I don't think they'd use their nukes. Too many old Cold Warriors with long memories in Moscow to open that nightmare again."

Aster shrugged and stifled a yawn. "You may be right, Admiral. I'm not sure about anything anymore." He checked his pilot's chronometer and excused himself.

Moments later, Lee and Lieutenant General Dave Forester were complicating Androsin's already mind-numbing day. The Army colonel had agreed readily to initiate a side-wargame for Lee, a focused effort on China, per the go-ahead from General Aster. But Forester, as director of operations for STRATCOM, had other priorities and was balking at providing support for it.

"Admiral, I know damn well the fire-breathing dragon could burn our butts, so your side-game on China makes good sense. I'm with you on that point. But I don't want my battle planners sucked into it, okay?" Forester stressed, shaking his head emphatically, arms crossed. "I want my folks one hundred percent focused on Iran right now. Intel is humming about those crazed Islamists preparing to launch another Shahab, and there's a very good chance it'll be carrying another nuke. Hell, the whole frickin' world is nuke-nuts all of a sudden! We *have* to be ready to head off the next nuke-happy idiot, or we'll all be crispy critters!"

Androsin had campaigned hard to get Forester's battle staff integrated with his wargamers as a way to shorten the ideas-to-implementation cycle. Aster had seen the potential payoff and green-lighted that integration. But Forester had been skeptical from the outset, and the latest flurry from North Korea was hardening his belief that his battle guys were wasting time in the wargaming center, valuable time better spent preparing real-world war contingencies.

"Sir, I read you loud and clear," Androsin nodded. "I guarantee we'll be working Iran and North Korea as top priorities. BOYD's crunching open-source and intel info, looking for a pattern that will dictate our next round. As soon as Jill's pulled that info together, we'll be back on Iran and Korea, and we'll shoot everything to your staff," Androsin assured him.

"Sounds good," Forester said. He glanced at Lee. "I'd appreciate getting updates on whatever you come up with concerning China, sir. Especially any PLA link to the North Korean nuke event, or to Iran. We sure as hell can't afford to be blindsided by a coordinated one-two punch from both sides of the world." He hesitated a beat, then locked eyes with Lee. "How in *the* hell did we get into this mess?"

The retired admiral shook his head and pointed at a large plasma display dominating the Center's front wall. It depicted a view of all satellites in orbit, as if one viewed the Earth from a perch high above the North Pole. Slowly orbiting icons were now colored red and blinking slowly. They represented spacecraft that had been designated "no longer operational" by the Joint Space Operations Center at Vandenberg Air Force Base, California. Even as Androsin, Lee, and Forester scanned the display, three more red blinkers appeared. Military, commercial, and civil satellites owned by more than half a dozen countries and myriad companies were either dead or dying. The North Korean blast had not been selective. Everybody's space platforms had taken a hit.

STRATCOM HEADQUARTERS/COMMANDER'S OFFICE

General Howard Aster's "B.S. flag" was waving. Via phone, the four-star in charge of U.S. Air Force Space Command, General Erik "Buzz" Sawyer, was trying to convince the STRATCOM commander that immediate action against a constellation of China's tiny "nanosatellites" was necessary. Aster was far from convinced.

"Damn it, Buzz! That doesn't make a lick of sense! Granted, I'm no lawyer, but taking out another country's satellites is tantamount to an act of war, as I see it," Aster argued. "Besides, thanks to Kim's high-altitude nuke, spacecraft in low-Earth orbit are dead or dying like flies in December, *just* from exposure to heavy-duty radiation. Why risk an international incident, or even worse, by going after those Chinese nanosats now? The little critters are probably already fried! Whacking them on-orbit would be pointless!"

"Howard, that's exactly the same argument I made last night, when my Fourteenth Air Force space cadets proposed this," Sawyer explained. "But my guys built a dad-gummed convincing case for *temporarily* disabling those Chinese nanosats, and I finally had to agree with 'em. The code-breaker, for me, was that 'temporary' part. We just need to buy time, a few weeks, at most. And that can be done very quietly, without sparking an international incident. We just poke out those little buggers' nano-eyes until either those two carrier groups are out of missile range, or we get China settled down a bit."

"C'mon, Buzz! What do you mean 'temporarily' shut 'em down?" Aster demanded. "I suppose your guys are gonna fly up there and smear soap on those tiny satellite lenses, then wash their little windshields later. And China won't even notice, right?" Aster's cutting sarcasm was legendary throughout the Air Force. On the flight line, the lighter version of that biting sarcasm, delivered with a lopsided grin, had often sparked a laugh from an F-16 crew chief or wingman. Of course, it hadn't played particularly well in Washington, during Senate hearings. Still, the former fighter jock had squeaked through the politically charged confirmation process and received both his fourth star and assignment as STRATCOM's commander.

Sawyer chuckled. "Well, sort of. Before I outline the 'how,' let me walk through the 'why,' all right?"

"Sure, sure. Press on," Aster growled. Now he was pacing, as well, following the pale-blue carpet's well-worn path around his desk and impressive slab of an oak conference table.

"First, my guys say there's no guarantee that China's nanosats *were* zapped by the North Korean nuke. In fact, it appears the volume of signal traffic with their nanosat constellation has actually *increased* since that blivet went off at high-altitude," Sawyer said. "The Chinese have gotten much better at hardening their spacecraft in recent years, and intel says they use nothing but rad-hard electronics components. . . ."

"While we've moved in exactly the opposite direction, relying more and more on off-the-shelf commercial components," Aster interjected. "Everything we've fielded since the early nineties is *not* hardened against EMP or intense radiation, because . . . what the hell? 'Nuclear war's a thing of the past, right?' Guess the inscrutable, clever Chinese didn't buy into *our* 'wisdom,' huh?"

"Damn straight. That's why those nanosats probably rode out the nuke blast and are still beepin' and squeakin'—at least for now," Sawyer hedged. "No way to tell when or if they might roll over and die, but today they have to be considered a threat. That's the way the Navy sees it, too. I don't blame 'em for being damned nervous!"

Aster stood at the window, staring at a windswept prairie stretching away from Offutt Air Force Base and STRATCOM headquarters. "No joy, Buzz. I still don't see a significant threat. Those nanosats watch and report, just like satellites have done for decades. We have dozens of 'em in orbit, doing the same thing. So does Russia, England, France, Japan, India and . . . hell, probably half a dozen other countries. Why, suddenly, are *these* considered threats and why *now*?"

"Several reasons, Howard. Based on serious characterization work our guys have done on those nanosats, we know the little critters are able to track our carrier groups in the northern Pacific, and can relay both imagery and very accurate position information for targeting. Those boats are sittin' ducks for Chinese missiles, thanks to those nanosats flyin' overhead. If China decides to go ballistic on us, you and I both know they'll be shootin' missiles at those carriers and probably sending fighters out to take potshots, as well. But if we pull the shades on their nanosat eyes-in-the-sky, the Chinese will have no idea where our boats are, and might think twice about shootin' in the blind."

Sawyer paused to ensure that idea registered with Aster. "And you know what happened on Speed's last spaceplane mission, Howard. Just because he flew up close to those damned nanosats, the Chinese Army fired lasers at him. The bird's canopy was weakened by a couple of laser shots, and came within a whisker of splittin' wide open during Speed's reentry! I'd say *that* constitutes an unprovoked act of war, too, doesn't it?"

"Hell, I don't know, Buzz," Aster said, sighing wearily. "I've spent a career flying below fifty thousand feet. If somebody shot at my airplane, there was no question whether I could shoot back or not. In space . . . Who knows? Our space-lawyers are still trying to sort out the legalities of what happened to Speed on that XOV flight. With no precedents, there's not much to go on, and *that's* giving the space legal-eagles a brain cramp.

"But you and I aren't going to resolve that one," Aster said, abruptly returning to the issue at hand. "Look, I'll consider Fourteenth's proposal to take out those nanosats, but only *temporarily*. Have your guys send their point-paper to my exec, and copy Dave Forester, my ops chief. He's pretty much living here full time, either in the command post or in our Deadsats II-Plus follow-on wargame. I want his input on this."

Sawyer nodded, then tried one more shot. "Roger that, sir. While you and Dave are mulling this over, though, I'd like to get folks at Vandenberg and Kirtland teed up. We can shut 'em down at any time, but prepping a rocket and that XSS-13 minisat for a short-notice mission will take awhile."

"Understand. Get ready to launch, but make sure your guys don't start mashing buttons until I give the order, copy?"

Sawyer smiled. "Yessir. I'll get back to you as soon as that Falcon at Vandenberg is ready to fire, then put everything on hold till you wave the green flag."

Once the two generals signed off, Aster folded his arms and stared through the window for several minutes, wrestling with the pros and cons of Sawyer's pitch. He didn't like the idea of dinking with any country's satellites, especially China's. The potential for getting caught in the act was high. Would China retaliate yet again? Tensions were already escalating in the Pacific, and this anti-nanosat mission had the makings of yet another round of "unintended consequences."

On the other hand, he couldn't afford to leave thousands of sailors on those two carrier task forces at risk, either. For whatever reason, Admiral Lee suddenly had a bug up his tail about China, too. Something had triggered that cagey old Navy bird's warning receiver, and Aster had immeasurable respect for Lee. *If things turn to mud, and China starts firing missiles, it'll be too damned late to smack their nanosats. And if those little birds help China zero-in their missiles and fighters* . . . Aster shivered slightly as images of burning ships and dead sailors flashed through his mind. He sure as hell didn't want to carry *that* burden the rest of his life.

Aster headed for the office door. He'd have Forester's battle staff take a hard look at what Buzz Sawyer was sending up from Colorado, then get their assessment. Forester was in a snit about his battle staff being tied down in a wargame, so this would give him something solid to sink his teeth into. But STRATCOM's top general had already made his decision. The XSS-13 mission was a go.

VANDENBERG AFB, CALIFORNIA

Standing on a long tongue of ragged-edged, white-hot flame, SpaceX's slender Falcon 9 rocket climbed vertically, its blunt nose piercing a thin cloud layer. A distinctive, deep-throated crackling echoed across fog-dampened, dark hillsides miles from the Vandenberg AFB launch pad, diminishing in volume as the booster accelerated. Its logo-covered body briefly caught the weak rays of a dawning Sun, then shrunk to a tiny bright spot tied to Earth by a slowly twisting rope of thick, white smoke hanging in the dark-blue California sky. The hybrid rocket and its mysterious payload were space-bound.

Minutes later, miles above a receding Earth, the rocket's bulbous nose cone split on schedule, exposing a squat, compact craft. Pyrotechnic releases fired, freeing the microsatellite payload from a booster's exhausted final stage. The latter would gradually slow, its velocity eroded by the drag of high-altitude, thin air, until the rocket body fell back into a thickening atmosphere and burned up.

Computers on the microsat activated sophisticated sensors that quickly scanned a dark sky and found a collection of stars that matched a specific constellation stored in digital memory banks. The sensors shot bursts of navigation data to other microprocessors, which, in turn, triggered brief puffs of gas from tiny divert rockets, reorienting the spacecraft.

The advanced space platform had been designed, developed, and tested by a consortium of military labs and small aerospace companies under a Defense Advanced Research Projects Agency-run program. Over several years, that DARPA-led consortium had developed and tested the technologies necessary to autonomously refuel and service satellites in-orbit. The successes of two pioneer R&D satellites on the Orbital Express rendezvous-and-servicing mission in 2006–2007 had given the U.S. a tremendous head start in close-proximity orbital operations. Subsequent research missions had refined those on-orbit maneuvers and procedures. Now, the lessons learned would be put to the test in a real-world "soft-kill" operation.

Through a parallel program, a now-disbanded Air Force Space Command Battle Laboratory team had quietly developed classified "counterspace" technologies for a novel "piggyback" mission on the XSS-13 microsat. XSS-13 was the latest in a series of experimental, highly maneuverable satellite-servicing vehicles the Air Force Research Lab had built over the past decade. Today, though, experimentation was secondary, deferring to a critical wartime mission in orbit. XSS-13 was about to log space-combat time.

STRATCOM's wargamers had fretted over ways to counter that small constellation of recently deployed Chinese nanosats that Brigadier General Hank "Speed" Griffin had photographed months earlier. Almost certainly, the nanosats had detected the presence of Griffin's XOV-2 spaceplane during its recon mission. Sensor data, relayed to Chinese scientists and People's Liberation Army officials, had led to the spaceplane being targeted by powerful Ground-Based Lasers.

Consequently, China's nanosats, the wargamers had concluded, constituted clear threats that should be neutralized immediately. A 14th Air Force colonel, via videoconference, had proposed using the XSS-13 to temporarily neutralize, but not kill, the nanosats. Such a "soft kill" would avoid handing the Chinese a pretext for measured retaliation or outright war, while also protecting U.S. interests and the Navy's carriers for a few critical weeks. "Neutralization" could buy time for the U.S., the wargamers had agreed.

Even before STRATCOM's General Aster gave the order to proceed, engineers and technicians were working around the clock to prepare the vehicle and its Falcon launcher for combat duty. Finally, the experimental XSS-13 spacecraft was in space and homing on China's imaging nanosats.

Designed as a low-observable, or stealthy spacecraft, the small XSS-13 slipped into orbit with the nanosat cluster, apparently undetected. Automatic-rendezvous sensors, algorithms, and divert thrusters eased the vehicle within inches of the first Chinese spacecraft, a hunter relentlessly stalking its unsuspecting space-prey from a blind side. A command uplinked to the XSS-13 from an Air Force Research Laboratory control center at Kirtland Air Force Base, New Mexico, commanded a flexible robotic arm to slowly extend, positioning a tiny machined nozzle near the edge of the nanosat's lens. By all indications, the Chinese satellite and its ground-based controllers had no inkling the XSS-13 microsat was in the neighborhood.

A USAF colonel in the control center stared at a surprisingly clear image of the XSS-13's robotic arm being crooked ever so slowly toward the first nanosat's lens.

"That's good. Let's do it," he commanded. An operator acknowledged the order, tapped a key, and watched a short burst of greenish material squirt from the heated nozzle. That "slime" adhered to the nanosat's optics and froze instantly, transforming to a thin, opaque coating. Deep-freeze cold, time and cosmic rays constantly bombarding that coating would slowly deteriorate it. Within weeks, the green "slime" would flake off, leaving the lens clean and restoring China's view from space. Only a few unexplained particles would remain, floating aimlessly around the spacecraft. For now, though, the nanosat was blinded, its Earth-staring lens covered in frozen, opaque gunk. The U.S. had just fired its first nonlethal, reversible shot of space warfare.

"Soft-kill Number One," the colonel announced firmly. "Folks, that's the first imaging satellite to be 'slimed' in orbit! Well done!"

A subdued round of hoots answered, but everybody knew there was still a lot of work to do before *real* self-congrats were in order. Any one of those nanosats might detect XSS-13 and sound the alarm, possibly triggering another blast from ground-based lasers in China. Although hard to detect, the stealthy American research satellite was not designed to soak up megajoules of laser energy. One zap and their multimillion-dollar baby would be another pile of dead space junk. And those carrier groups would again be subject to very accurate missile attacks, thanks to data supplied by the remaining, still-functioning nanosats.

The USAF operator tapped another set of keys, commanding the XSS-13 to automatically fly toward a second nanosat, again approaching from the tiny Chinese satellite's blind side. Engineers in the control center scanned a stream of data being telemetered from the U.S. microsat, then gave "go" signals, in turn, when queried. While his colleagues held their collective breath, the operator carefully extended XSS-13's flexible arm a second time, squirted a measured burst of "slime," backed away, and started maneuvering toward the third nanosat.

In short order, the entire formation of tiny Chinese spacecraft was rendered temporarily blind. Additional commands sent the XSS-13 into a new, higher orbit, simply another experimental platform running on autopilot. With luck, nobody on Earth had tracked its stealthy dance among China's nanosat array.

The USAF officer-operator who had maneuvered the XSS-13 slumped, finally able to relax. His olive-drab flight suit was sweat-stained, a measure of stress, not physical activity. His commander, the stern-faced colonel hovering near his subordinate's shoulder throughout the tense engagement, clapped the young man on the back.

"Congratulations, Lieutenant. You've just become the first 'soft-ace' in space!" The colonel forced a half-smile, absently wondering whether that shallow distinction would matter, if China responded aggressively to the U.S. "slime attack." Many in the room feared the "Dragon" *would* retaliate. How and when was anybody's guess, though.

4 WHEAT AND CHAFF

STRATCOM WARGAMING CENTER

Effects stemming from the North Korean nuclear blast ranged far beyond material impacts on spacecraft and terrestrial computer-based systems. There were deep impacts on people, as well. Their day-to-day thinking and attitudes were affected, forever altered by the shocking magnitude of yet another nuclear event. People across the globe suffered from an unsettling, always-present anxiety, a collective angst fueled by the stark realization that modern life was incredibly fragile and tenuous. Not since the early days of the Cold War had so many felt so vulnerable.

Even the experts assembled in STRATCOM's wargaming center were subdued. Iran, China, and now North Korea all factored into an increasingly perilous outlook for America and the civilized world. The loss of "situational awareness"—that all-important knowledge of others' activities—made hard-line factions very dangerous players in an unfolding high-stakes drama.

Admiral Lee scanned the wargaming center, looking for STRATCOM's Director of Operations, Lieutenant General Dave Forester. He found the three-star Army officer huddled with his staff, reviewing the latest BOYD assessments forwarded by U.S. combatant

commanders. Forester's battle-planning team was often at the leading edge of national tensions, and Forester, as commander of the battle staff, was *the* spear's tip. The Army general was wearing down, a
fact that increasingly concerned the retired admiral. Forester was too
important to lose.

Before Lee could get the three-star's attention, Jill Bock shouted,
"Hey, boss! BOYD is spewing a ton of new open-source info! And
it's *most* interesting!" Forester and Lee turned toward the analyst.

"In a second, Jill," her boss, Colonel Androsin, replied, preoccu-
pied.

"Damn it, Jim! Listen up a minute!" Jill barked. "Remember when
I briefed General Aster about the 'GloCon' issue? How we were getting GloCon feeds from some of our own people?"

Lee glanced at Forester. "GloCon?"

"New form of what used to be called a 'blog,' sir," the general explained. "They're sort of advanced commentaries on niche subjects."

"So . . . what does 'GloCon' mean?" the retired officer pressed.
"Sounds like a radioactive felon."

Forester half-grinned, his tired eyes softening. "Short for 'global
conscience.'"

Lee shook his head. "Good grief . . ."

Nearby, Androsin waited impatiently for the BOYD update Jill
was formulating. *Impatience is becoming an all too familiar commodity around here,* he observed. Impatience, fatigue, testiness.
All were becoming palpable throughout the wargaming center. Fatigue was the most worrisome. Sleep had come as brief, irregular
snapshots for the entire headquarters staff in recent months. Personnel of all ranks—and even the wargame's civilian guests—were
feeling tremendous strain.

"Jill! You were saying . . . ?" he prodded.

"Colonel, BOYD is buzzing about China," she answered.
"BOYD's still doing some sorting and categorizing, but . . . Bottom
line: PLA has a lot of stuff on the move. Not just command-and-
control signals, but no-kiddin' maneuver units prepping to move.
Bombers, army divisions, subs—you name it." She was in characteristic, rapid-fire delivery mode, fingers flying across the keyboard

as she did what she did best: coax correlated, fused information from the BOYD computer-and-network complex.

"Jill, hold that thought," Androsin ordered, raising a palm. "Here's another intel report from the Defense Intel Agency. They're saying it's fairly *quiet* on the China front! Only the ongoing fuss over the North's nuke blast."

"Yeah, of *course*, there's no intel community updates!" she shot back, irritated. "That's 'cause they're waiting for collection of targets, then they'll go through all their anal analysis drills, then they'll staff their intent messages—making sure there's consensus all along the way."

Androsin was leaning over Bock's shoulder, scanning her workstation screen. "Yeah, but how do we separate the fly doo-doo from the pepper? Good info from nonsense?"

"No sweat. BOYD's discounting a deception campaign, because of the disparity of sources. It either dovetails, or an input gets trashed. See?"

Androsin straightened and suppressed a yawn. "Okay, Jill. Pull up that BOYD summary and let's get it to General Aster. *Now!* We're already late with an update, and he'll be screaming for it soon."

Androsin and Jill were soon outside Aster's office, on ice until the general finished perusing a summary of the Pentagon's latest tasking: *Remain focused on Iran. Develop deterrence strategies. At a minimum, consider global-strike options to prevent another launch,* the document ordered. Aster shook his head, tossed the sheet into a red-striped folder marked TOP SECRET, and waved to Androsin and his BOYD expert, Jill Bock. It was getting to the point that he dreaded hearing their latest wargame summaries. Good news was in short supply.

Back in the STRATCOM Wargaming Center, Admiral Stanton Lee was briefing a small cell of wargamers, explaining the guidelines for a new, China-focused sideline wargame.

"Key on the competing power influences within China. Explore Chinese courses of action regarding North Korea, and the United States' ratcheting up of military operations in the Pacific," the former naval officer directed. Lee sighed and added, "And certainly, Taiwan uncertainties."

Taiwan. Always Taiwan, he noted silently. Lee suggested the "China cell's" members should get started, excused himself, and walked to the center's coffee bar. He poured a fresh cup, then found a quiet corner. He needed to think, go over all the elements related to China.

Instead, his mind soon wandered, returning to a simpler time. He reflected on the thrill of dawn launches in an F-4B Phantom II, and later, the F-14C Tomcat. In his mind's eye, Lee again watched the huge aircraft carrier turning into the wind, the rim of a new day's sun slipping above the carrier's stern, as scores of technicians in multi-colored jerseys swarmed over the flight deck. *Just memories now. What you think will last forever is ultimately so damned fleeting!*

A flood of love for his wife suddenly coursed through him. *And what of our future?* He felt an intense, burning desire to bail out of the center, to simply go home, to spend the rest of his days with her, walking together, holding her hand. . . .

"Admiral? Admiral, sir?" a young Air Force captain's soft, but persistent call drew Lee back to the present. "General Aster has requested your presence in his office, sir."

"Thank you, son. I'll be with the general shortly." Lee rose with effort. *You can't know the sea by standing on the shore. You can't know China by standing on America's shores, either. What in the hell were we thinking for all those years? Containment? Engagement? Both looked the same to the Chinese. It was really all about containment; that's what China believes, anyway. And that's why they embarked on their own military transformation.*

For more than twenty years, China had pursued asymmetric strategies and technologies aimed at conducting and winning an eventual war in space and the new, digital realm of cyberspace. It began modernizing its Navy, transforming for power projection, as well as to defend its shores. Each of those pursuits reflected its leaders' adamant belief that America's long-term goal was actually very simple: to contain China. In parallel, the Chinese had quietly resumed their centuries-old campaign of imperial expansion.

Oh, its leaders wouldn't admit that publicly, but modern China firmly believed it was in a protracted struggle with Asian neighbors, as well as with the West, over control of limited natural resources. From its perspective, the world's supply of raw materials was finite. Responses to global warming threatened to throttle growth, and industrialized economies were locked in a vortex of uncertain futures through competition for markets.

That competition was a zero-sum game, as the Chinese saw it, but a still-powerful Communist legacy clouded rational thought about the impacts of unchecked industrialization. China's air pollution had been an acute embarrassment during the 2008 Olympics, but, since then, it had become deadly, killing its citizens at an alarming pace. The spread of AIDS was virtually unimpeded in the western provinces. Shanghai and Hong Kong had competed for the best business and technical minds, but the outcome diverged from Beijing's grand plan. The two enclaves of capitalism and liberal thinking had become hooked on the self-indulgent drug of material pleasures. And throughout the nation, crime had become rampant, corroding confidence and trust everywhere.

Consequently, increasingly isolated Chinese leaders held fast to their "old-think" ways. They began to exhibit the irreversible, steady fault lines of a failing regime. They needed an enemy to keep their increasingly disaffected citizens diverted from the harsh realities of daily life. Among the more-militant circles of power, America had become that enemy.

We took our eye off the ball, after the Cold War ended. Lee grimaced. *Now, we're dealing with the results of that benign neglect. Damn!*

Entering Aster's outer office, Lee greeted Annie warmly. She turned from her computer, removing her reading glasses, and shot him a warm smile.

"Admiral Lee! Please come in. The general is expecting you," she said, rising. "May I get you a glass of iced tea? I believe you take that with lemon?"

"I certainly would appreciate that, Annie," he smiled over his shoulder as he entered Aster's office. *Such a sweet lady*, he reflected. Although women had never been a part of *his* world-at-sea, he appreciated their warmth and uncanny instincts.

"Hello, Admiral. Thanks for coming up. I know you're damned busy," Aster greeted Lee, waving him to an overstuffed chair. The STRATCOM chief finished signing something and handed it to Burner Burns, his military aide, then joined Lee as the lieutenant colonel departed, closing the office door.

"We're all busy, General. There's a war on, you know." Lee's half-hearted attempt at dry humor fell flat. "What can I do for you?"

"Do you know a Doctor Lin Zhang?" Aster asked, leaning forward, elbows on his knees.

"I do. A senior fellow of the Shanghai Institute for International Studies. Thoughtful guy; very influential. He's considered a moderate, although he's written some fairly hard-line things about our intentions in Asia, particularly with respect to China. Why do you ask?"

"Admiral, what if I told you there was a back channel opening up, giving us a direct link to top Chinese leaders? The contact is Zhang," Aster said.

Lee reflected for a moment, nodding absently. "I'd say that's fairly true to form. China's government officials *would* opt for a more-indirect approach. Yes. It makes sense. Save face, especially in a time of crisis. Such initiatives usually come from a particular faction, however. Always covertly, even within the highest reaches."

"Precisely the situation here," Aster nodded. "The President's chief of staff just forwarded a highly cryptic, extremely close-hold message to me. He even dictated the damned thing. Didn't want to send it any other way, he said. It seems our guy, Zhang, laid down some strong terms about disclosure, too. So let's just say that this has *not* been coordinated with the National Security Advisor."

Lee made no attempt to hide disdain. "That's a good move, Howard. Paul Vandergrift is a dangerous man."

Smiling, Aster agreed. "Yeah, you might say that! Hurlburt ripped him a new one over that Sting Ray 'worm' fiasco, to say nothing of getting Charlotte Adkins executed! The guy's an idiot *and* a dangerous butt-head!"

———

"So, what does Zhang propose?" Lee asked.

Aster extended a note he'd written, during the phone conversation with White House Chief-of-Staff Gilbert Vega. "That's exactly the way it was read to me. Seems *you're* the guy the Chinese want to talk to."

Without looking at the note, Lee protested, "Surely not . . ."

"Whatever. The President wants you back at the White House, ASAP. He wants your firsthand, private counsel on this matter. And I suspect he'll send you to China." Aster studied Lee, then flicked a hand. "Go ahead; read it."

Lee took the note and settled into the soft chair, surprised at this turn. He read each word with care, searching for the hidden or double meanings he knew would be there.

> *From the shores of great seas, one can almost see the eagle circling its prey. From these shores, I see the eagle as a fierce, but lonely hunter. What do you see from your shores? An angry sea? Or, if one moves farther down the shore, is there calm? Will the good Admiral Lee wear the shoes of a fisherman, as only a fisherman knows the sea? The admiral must be the only link to the President.—Zhang.*

'The shoes of a fisherman,' Lee thought, taken aback. *Could it be . . . ?*

5 SLEEPER CELL

TEHRAN, IRAN

Hassan Rafjani closed the door to his simple, sparse office. Its walls were devoid of photos or artwork, and the small, scarred desktop was empty. A thin, brushed-aluminum notebook computer was the sole occupant of a typing table extending perpendicularly from the desk. A wooden chair scraped across the stone floor as the gaunt-faced man pulled it aside.

He tapped a key, bringing the computer screen to life. His trembling hands betrayed a nervous excitement. Finally! After years of preparation, planning, maneuvering, and sacrifice, the moment had arrived. He looked at his fingers, poised over the computer keyboard, impatiently waiting for the device's "wake-up" cycle to complete.

Those two hands now held tremendous power. That power would reshape the Middle East and return Persia to its rightful place in the universe, he reflected, savoring the moment. And the transformation would begin at this very moment, when those hands typed a simple message.

The irony of his employing an ultramodern, electronic computing-and-communications machine to trigger a new holocaust, a cleansing fire that would return a depraved world to the simple, pure Islamic lifestyle of the twelfth century, was not lost on the well-educated Rafjani. He pecked a cryptic message, checked it against a stored file, then posted it to a special Internet Web site.

It has begun, he thought, immensely satisfied. He felt strangely at peace, yet simultaneously tingling with excitement. A wailing call to prayers pierced the dirty window placed high on the wall of his room, pulling him back to the now. He quickly verified that the brief message had appeared on the correct Web site, then signed off and lowered the computer's screen. Smiling, he stepped to the threadbare rug in a corner and knelt. Allah would be most pleased today.

APARTMENT 911/POMONA, CALIFORNIA

Khalid was ecstatic. *Finally! A mission worthy of Allah's Soldiers!* He reread the obscure posting on a local community college Web site, relishing the words that released his brothers and him to action: *"Wanted: Family-oriented man of Arab descent to share life with a traditional Muslim woman. . . ."*

The rest was fluff, a carefully constructed electronic missive designed to deflect Satan's FBI dogs from the many small groups of Muslim operatives hidden among American heathens throughout the United States. Khalid had waited many long months for this innocuous personal ad to appear on the college's chatty Web site, searching in vain every morning for a specific, coded message posted by the Master, a high-level religious and political official in Tehran.

And today was *the* day he, Khalid, had been commanded to unleash Allah's fury on the infidels of Southern California! And, oh, how he looked forward to bringing death and destruction to the smug, immoral, self-centered Americans—especially those tanned, half-naked beings who chose to live like rats in the crowded warrens of greater Los Angeles, battling each other on the concrete freeways and broken-asphalt streets that despoiled God's holy Earth.

"Soon, the infidels will die a horrific death . . . God willing," he whispered, meticulously closing all the application windows on his laptop computer's screen. Even if the perverted FBI law-dogs invaded his tiny apartment—an abode he'd selected because the number on its door was revered among those called to serve Allah's army—they'd never find a scrap of information that could link him to the devastating events about to materialize. He smiled, already thinking ahead to his mission, mentally rehearsing each step.

Khalid and his five well-educated and -trained Iranian and Syrian compatriots had slipped across the Mexico-California border almost two years earlier. Even though they had planned the crossing meticulously, then paid a Mexican "coyote" to physically transport them, their entry to the U.S. had been embarrassingly simple. No border guards, no high-tech sensors, no barrier of concern. Within a few hours of slipping across the cleared and fenced no-man's-land that separated the two countries, Khalid and his fellow jihadists were sleeping in warm beds, miles inside California. That experience had only bolstered his contempt for soft, mindless Americans, and deepened his belief that Allah, indeed, wanted pure-of-heart Muslims to eradicate the immoral scourge that America had become.

He carefully slipped the laptop computer into a shoulder bag and set it aside. His heart dancing with joy, he unscrewed the back cover of an old-fashioned, wooden stereo cabinet, exposing its electronic contents. But the system nestled inside was anything but a dusty, 1970s-vintage music system. He lovingly removed a carbon-composite chassis encrusted with tightly packed microcircuits, a modern, compact power supply and an intricately folded antenna. This system was the product of countless hours spent acquiring components from RadioShack and his employer's stock room, then painstakingly assembling them on his apartment's tiny kitchen table. The sophisticated system had been meticulously tested, ensuring it would perform precisely as the Master had dictated.

Before the day is out, America will again feel the bite of Allah's singing sword, he mused, a thought that excited Khalid tremendously. But there was much to do, before his electronic "sword" could sing. His fellow jihad soldiers would already be moving, preparing a complementary part of the operation in their Long Beach high-rise, towering above California's coastline. He must be ready to strike at the same time, or the impact would be unacceptably softened.

Alone, miles inland, Khalid would act from a surveyed perch in the mountains east of Ontario. Although his mission was less complex, it was every bit as important as that of the Long Beach team. If both were successful, Southern California's millions would be shocked, horrified, and terrified. And the huge state's booming economy would screech to a halt. Body counts would not be as high as those tallied by al Quaeda's 9/11 heroes, but his and the fellow jihadists' one-two punch would take a tremendous toll on California's economy, and ensure the state's people would never feel safe again. The Master had emphasized that striking one of America's key economic centers was a far more worthy goal for *his* highly educated, well-trained jihadists. Besides, the time for shocking, impressive body counts had passed, giving way to sophisticated, hard-hitting blows that would stagger the infidels' corrupt society. Knocking out dozens of satellites was the first step, a critical element that made the next phases possible. Obviously, the Master had succeeded in that mission.

Khalid and his team were now free to attack. They were Hezbollah. *We cannot, will not fail—God willing.*

COCKPIT OF ARCADIA AIR BOEING 757, FLIGHT 623

Captain Bert Kiley watched the top-of-descent marker march downward, toward the tiny airplane symbol on his navigation display. His right hand rested easily on the Boeing 757 airliner's dual throttle levers, prepared to pull both aft, as soon as the marker and aircraft-like icon merged. The casual, yet purposeful motions bespoke his many years on the 757's flight deck. As a junior pilot, Kiley had transitioned to the airliner shortly after it left Boeing's factory in the 1980s. He loved the big bird, and never moved to another type. Admittedly, both the 757 and he were getting a bit long in the tooth, but Kiley figured he'd make it to the FAA's age-65 mandatory retirement, before Arcadia traded its 757s for a newer model.

"Any change in the weather?" he asked the pilot to his right.

"Not yet. Ontario's still showing a fifty-foot ceiling and half-mile visibility in heavy fog," First Officer Kevin Sperling responded. "Multiple, solid cloud decks below 15,000. We'll be in the goo all the way down."

Sperling tapped a keyboard on the button- and knob-cluttered console that separated his and Kiley's sheepskin-covered seats on the air transport's roomy flight deck. He scanned a datalink message transmitted by Arcadia Air's inflight communications provider, Aeronautical Radio, Inc., before adding, "Fog's spotty, though. Light winds clear the runway now and then, so a few guys are gettin' in okay—at least to Runway Two Six Left."

Kiley grunted, noting that his flat-panel display's tiny aircraft symbol now touched the top-of-descent icon. He punched the Vertical Navigation, or VNAV button, on the "eyebrow" instrument panel in front of him, and felt the "autothrottle" system smoothly retard the twin levers under his right hand, commanding the 757's massive turbofan engines to wind down. They settled to a constant, low-pitched moan as Kiley tweaked the throttles to a specific N1, the engines' fan-speed readout, a few percent above idle. The autopilot eased his big airplane's nose over to a predetermined descent angle. Several wiggles of the throttles satisfied Kiley that his airliner had "captured" the proper descent path. Later, he'd adjust flaps and gear and set up for a landing at Ontario International Airport.

He glanced outside the wraparound windscreen, checking that the pattern of tonight's bright stars had stabilized, affirming the 757 was at the correct angle with a faint horizon.

One peek beats a hundred cross-checks, he smiled. For some reason, the old instrument-flying maxim had flashed across his mind. Although professional pilots were adept at controlling an airplane solely by reference to instruments, a set of human eyeballs was still the best tool in any cockpit.

Descending through 30,000 feet and still more than a hundred miles from the airport, the pilots could barely discern a faint demarcation between star-studded sky and a jagged, black horizon, where air and ground met. Out there, where they would land, he could see absolutely no lights tonight. Usually, the greater Los Angeles basin was an endless sea of twinkling lights just beyond the dark silhouettes of mountain peaks that lay between them and their destination.

Per the airline's standard procedures, Kiley and Sperling verbally reviewed the landing approaches to either of Ontario, California's, dual east-west runways. Nestled two miles east of the sprawling city, the airport was uncomfortably close to the sharply rising San Gabriel Mountains to the north. Its runways paralleled the mountains' ridgelines, so that wall of rising terrain didn't worry Kiley. But, before they got close to the airport, they'd have to thread the airliner between two massive mountain ranges bracketing a narrow corridor called the Banning Pass. He'd never liked having to shoot that gap. Especially at night and in thick clouds, when he couldn't see the deadly rocks on those mountains.

Tonight, the San Gorgonio Wilderness mountains would be on his right and the San Jacinto Wilderness on his left. Gorgonio's rugged terrain reared up to 11,502 feet, the highest in the Los Angeles area. San Jacinto's topped out at just under 11,000 feet. But the Ontario airport was farther west and far below the peaks, registering only 944 feet above sea level. Aircraft landing at Ontario from the east literally dived toward its runways, after navigating the terrain of Banning Pass. Tonight, with those cloud-covered, rugged mountains rising on each side of the pass, the pilots were completely reliant on aircraft instruments and navigation systems to ensure they stayed on the correct flight track. Assuming everything worked as planned, Kiley would level off as his 757 emerged from the Banning Pass, pick up the invisible, electronic beam of the Ontario Instrument Landing System, or ILS, and follow its three-degree glide slope to a tire-chirping touchdown.

"How's GPS looking?" Kiley asked, eyes still scanning several large, flat-panel screens filled with numbers, a prominent magenta-colored, curving flight path and other data.

"Still shaky. 'Bout the same as it's been the last few days."

"Damn!" Kiley swore. "How long will we be dealing with this screwed-up ionosphere?" Sperling didn't answer. He focused on typing commands into the Flight Management Computer, essentially the navigation brain of the Boeing 757's automated guidance system.

That FMC, coupled with a sophisticated autopilot, would guide their 757 down into the safe haven of Banning Pass—and well clear of those looming mountains.

A few days ago, the FAA had notified all U.S. and foreign airline flight operations departments that a North Korean nuclear weapon blast at high altitude had literally charged up the ionosphere, essentially creating a persistent electronic shield around much of the Earth. Consequently, faint position and timing signals beamed down from the constellation of twenty-four active Global Positioning System satellites, all orbiting 12,500 miles above the globe, were sometimes blocked by the irregularly charged ionosphere. It would dissipate, but not soon.

"Could be months, according to the feds," Sperling finally answered. "Nobody knows for sure." He was double-checking the FMC programming, ensuring memory-stored coordinates for the approach end of Ontario's Runway Two Six Left were correct. He'd seen way too many documented cases of improperly programmed FMCs driving pilots into the ground, including one horrible accident at his own airline. A single incorrect number or letter could spell the difference between life and death. It was Sperling's job to make sure those numbers were absolutely spot-on correct.

Kiley shook his head in disgust, but didn't reply. *Sure as hell. Just like we warned 'em!* A chorus of "old heads," pilots who collectively had logged hundreds of thousands of flight hours, had objected loud and long a few years ago, when the FAA announced the phased decommissioning of VORs, VORTACs, and other fixed-base navigational aids across the U.S. For decades, those stations had transmitted radio beams that pilots used to navigate the nation's airspace. They literally constituted invisible electronic highways in the sky.

But those stations were expensive to maintain. Confidence in the satellite-based GPS network grew as pilots routinely turned to onboard GPS receivers, which made efficient, point-to-point navigation possible. In the mid-2000s, the FAA decided to go all-GPS for U.S. air navigation, relegating the fixed-base navaids to aeronautical history.

Kiley and other silver-haired airline captains questioned that decision. Without ground-based navaids, they warned, pilots would be left with no backup to the satellite-based system. They asked, "What happens if we lose GPS signals, for some reason?"

The government's answer was less than heartwarming: Don't worry about it. GPS will always be there. It's *much* more reliable than the hundreds of white, inverted cone-shaped VOR and VOR-TAC stations that once dotted the national landscape. Besides, the FAA engineers noted, all modern airliners were equipped with companion inertial navigation systems. If GPS signals were ever lost—a highly unlikely scenario—the gyroscope-based INS units would ensure a safe arrival at the transport's destination.

"You relied on INS for over-ocean navigation for years, so what's the problem with relying on it as a backup to GPS over land now?" they asked. *Easy for you to say, ground hog,* Kiley reflected silently.

The FAA's sanguine approach did little to satisfy the most-experienced pilots, who had seen fancy electronics fail umpteen times during their careers. Two issues concerned Kiley and his fellow "graybeard" pilots: One, several GPS satellites had been attacked in recent months, causing four or five of the on-orbit birds to spew ever-increasingly inaccurate positioning signals. To their credit, the Pentagon and the FAA had acted quickly, telling airline operators how to reprogram Flight Management Computers, ensuring the "bad" satellites were ignored.

Two, the FAA's boffins had never adequately addressed a little-known issue: INS units were automatically relegated to minor-player status in the latest generation of sophisticated Flight Management Systems. Today's FMS computers simply used INS outputs as an "adjunct voting member" to augment GPS signals. And GPS data typically were given priority votes. If INS outputs and GPS information differed, GPS won, and the airplane's autopilot opted to follow the latter's direction.

Kiley tried to mentally sort through the technical hows, whats, and wherefores of GPS-versus-INS, but gave up. He was no engineer, and trying to noodle all that voting stuff gave him a headache. It was almost three o'clock in the morning, and, after a long flight from Orlando, his brain was late-night fuzzy. He sighed, stretched, and returned his full attention to the rapidly decreasing altitude reading.

"Captain, look at that," Sperling said sharply, pointing to an FMC screen's display. Kiley glanced down.

"What am I looking for?" he growled, turning back to the altitude readout: passing through 15,000 feet, descending.

"We've got a solid GPS lock," Sperling exulted, tapping a few keys on the FMC keyboard. "Damn strong signal, in fact. Maybe we're picking up Ontario's WAAS already. Looks like we have GPS guidance, for a change!" Sperling's sense of relief was obvious. Solid, reliable GPS signals meant the 757 could shoot the Banning Pass precisely on track.

To ensure extremely accurate GPS signals at key airports, enabling routine GPS-guided landings, the FAA had installed Wide Area Augmentation Systems, or WAAS, units at certain airfields. Precisely surveyed, ground-based WAAS sites received signals from several in-view GPS satellites, and automatically compared the GPS-derived position and timing information with the WAAS station's known, actual location. If there were differences, WAAS calculated a correction, then broadcast that "delta-position" correction information to all aircraft in the area.

In effect, the WAAS said, "GPS, you say my station's position is over *there*, but I know I'm actually *here*. Therefore, all you airplanes out there, crank in a 10.2-foot correction," or whatever number the WAAS computer said it should be. That correction was automatically fed into the onboard flight management system, making sure an aircraft's GPS-supplied position data were highly accurate, virtually guaranteeing the airplane was exactly where the onboard FMS and instruments said it was.

Tonight, unfortunately, that was a false guarantee.

MOUNTAINSIDE/EAST OF ONTARIO AIRPORT

Khalid shivered and drew a thin coat tighter around his bony shoulders. Slightly built, he wore a polyester jacket, a thick Cal Poly sweatshirt, blue Levi's jeans, and cheap running shoes. His eyes were locked on the glowing screen of a notebook computer balanced on knobby, quivering knees. Seated on a folding canvas stool, Khalid shared the camper shell-covered bed of a silver Toyota Tundra pickup truck with the electronic equipment he'd pulled from that old stereo system. A bank of wet-cell Delco batteries fed direct-current power to the sophisticated electronic system. His late-model truck was parked on a rough, two-wheel dirt track that zigzagged up the side of a brush-dotted mountainside. He was miles from the nearest structure, a dilapidated gas station that was only open during daylight hours.

Khalid checked the radium dial of his Pulsar wristwatch. Three in the morning. Time to check in. He pulled a cell phone from the jacket's pocket and hit a preprogrammed code. The call was answered on its first ring.

"Yes?" the voice answered.

"Hey, dude. I'm headed for work, so sorry for calling this early," Khalid said, trying to affect a cool Southern California surfer accent. "Are we still on for the game tonight?"

"No problem. I was up early, studying for an exam I gotta ace this morning. Yeah, I'm on for tonight. You good to go?"

"Yeah. 'Bout six-thirty, at the usual place?" Khalid asked. He concentrated on affecting a bored, twenty-something California persona, but sensed his words had an uncontrollable tightness to them.

"Right. We'll grab a beer and burger before going in. Lot cheaper than ballpark prices," the voice said. It, too, was struggling to appear casual.

"Okay. Gotta go. See ya tonight."

A faked yawn from Long Beach answered. "Have a good one." A click ended the call.

Khalid rubbed his hands, excited. The conversation's real message was anything but casual. The affirmation that "we're on" for tonight's L.A. Dodgers' baseball game meant his fellow jihadists were also in position, radiating bogus signals over Long Beach Harbor, just as Khalid was transmitting from his lonely, fog-shrouded "campsite" above Banning Pass.

He rechecked the computer's data, satisfied that his homebuilt system was working properly. An odd, spiral-shaped antenna temporarily mounted on the roof of his pickup's camper shell was, indeed, firing a narrow beam of digital ones and zeros up and to the east, into a cone of airspace transited by airplanes landing at Ontario International.

Khalid switched to another open window on the screen and confirmed that his wireless connection was still active. *Great!* He was hooked to the Internet, and the signal was rock solid. His "Flight Tracker" software confirmed that Arcadia Airlines' Flight 623 was on time. A few key taps revealed it was a Boeing 757, descending toward the Ontario airport. Khalid bared perfect, bright-white teeth in a broad smile, a sharp contrast to his smooth, caramel-colored skin. Flight 623 was already in his equipment's beam, hopefully swallowing bogus GPS "corrections" his unit was transmitting. But instead of removing GPS errors, Khalid's WAAS signals were inserting ever-increasing errors into Arcadia Flight 623's FMS, steadily coaxing the airliner off-track.

In military electronic-warfare terms, the young jihadist's relatively simple electronic transmitter was "spoofing" America's high-tech Global Positioning System. "Spoofing" was the art of tricking sophisticated systems, inserting bad data or commands, but without the knowledge of their human operators or being detected by the victimized systems' silicon watchdog circuits. U.S. Air Force "Space Aggressor" units, which played the role of enemy forces during training exercises, had proven that smart engineers and technicians could build a relatively simple system to spoof GPS and communication satellites. Measures had been taken to protect military equipment and train soldiers, sailors, and airmen how to detect and counter spoofing during the subsequent eighteen years.

However, commercial navigation systems, such as those used by airliners, remained highly susceptible to GPS-spoofing and -jamming. Enslaved to brutal cost-cutting pressures of the marketplace, airline managers had resisted incorporating such necessary, but pricey, technical measures, despite known vulnerabilities to GPS spoofing. As Captain Kiley had once told an Arcadia vice president angrily, "Hell, we can *never* afford to fix known problems—until a bunch of people get killed and the feds force us to! That's why FAA regulations are written in blood!"

Consequently, instead of diving into the open air above Interstate 10, which ran through Banning Pass, an Arcadia Airlines Boeing 757 was offset to the north, its GPS-based navigation system lying to its pilots.

ONBOARD ARCADIA FLIGHT 623

First Officer Kevin Sperling was mentally wrestling with a dilemma. Should he tell the taciturn Captain Kiley or not? Sperling again checked the glowing FMC screen to his left. No doubt about it. The inertial navigation unit disagreed with the GPS-derived position. Not by a few seconds-of-a-degree, either; their position readouts were several minutes apart—a matter of *miles* difference! But which was correct? The INS-derived position claimed the 757 was well to the north of Banning Pass, descending at more than 400 knots ground speed toward an 11,502-foot mountain.

But inertial nav systems, which relied on gyroscopes, either tiny rotating wheels or compact fiber-optic-and-laser packages, would drift over a period of time. That's why the archaic things were "slaved" to the GPS system, providing an automatic means to continually "update" the INS position and keep the two in sync. But that wasn't happening now. *Why?* He mentally worked through the FMS logic, trying to find a good reason for the different readouts.

Airline policy dictated that Sperling alert Captain Kiley of the INS-GPS discrepancy, but he hesitated. Kiley was the quintessential old fogey pilot. He had an inherent mistrust of anything modern and different, always bitching about new procedures and expounding endlessly about "how we used to do it." Old ways were always better than any cutting-edge new system or procedure, according to the grouchy captain.

Sperling stared straight ahead through the windscreen, toying with the pros and cons of telling Kiley. He should—but it wouldn't make one damned bit of difference anyway. There was really no choice. Ultimately, Kiley would go with the GPS guidance. No pilot in his right mind would believe a drift-prone INS instead of the flawless GPS-derived position and track information, so why bother going through a senseless debate at three-frickin'-o'clock in the morning? Still, the rules were as old as air-navigation itself. The junior pilot was required to alert his captain about nav problems.

"Hey, Captain. There's a discrepancy between. . . ."

"Through twelve for eleven-thousand," Kiley interrupted matter-of-factly, calling out the descent altitude. His eyes remained fixed on the flat-panel instruments. "Gimme a . . ."

His command was drowned out by a loud whoop-whoop aural alarm and a computer-generated voice frantically ordering, *"Terrain! Pull up! Pull up!"* The aircraft's ground-proximity warning system had detected rising terrain ahead.

Sperling simultaneously jerked upright and grabbed the control wheel in front of him. *"Shit! Pull up!"* he screamed, yanking the yoke into his lap, commanding an immediate climb. A gap in the gray, featureless clouds had offered a glimpse of pure pilot-hell approaching at what felt like warp speed. A faint glimmer of light from somewhere—aircraft wingtip position lights, maybe a yard light near some remote cabin—had reflected off a wispy cloud, dimly revealing a jumbled-rock hillside. A silly, fleeting thought flashed through the first officer's brain: *Why didn't 'Bitchin' Betty' sound off sooner?*

Sperling's scream and the control wheel being yanked out of his left hand were the last impressions that registered in Captain Bert Kiley's brain before the Boeing 757 smashed into Mount San Gorgonio's unyielding tons of rock and dirt. In a fraction of a second, the blunt-nosed airplane crumpled like cheap aluminum foil, spewing Jet A fuel from ruptured tanks, then exploding in a black-and-orange fireball. Fifty-seven passengers and crew members were smashed and shredded beyond recognition, dying instantly. All but the airplane's two pilots perished without a hint that they were crossing the fragile line between life and death.

MOUNTAINSIDE/EAST OF ONTARIO AIRPORT

Khalid held his breath, unblinking eyes locked on the computer screen's last-known position of Arcadia Flight 623. *It's gone!* he finally decided. A slowly crawling icon and data tag that had represented the Boeing 757 simply disappeared. Khalid had heard nothing, felt nothing, but knew the giant twin-engined airliner had crashed into the same range of mountains in which his Toyota was parked. Exactly where, he didn't know, but the airplane's transponder had stopped transmitting. Of that, he was certain.

It crashed! Just as the Master said it would! God is great! Khalid exulted. He leaped from the squat camper's back door and dropped to his knees, kissing the wet, fog-dampened grass. The Earth smelled simultaneously musty and deliciously sweet. On his knees, the young man reared back, eyes raised to the dark, fog-shrouded heavens and spread his arms wide, thanking Allah with all his heart.

U.S. NORTHERN COMMAND/OPERATIONS CENTER

"Ma'am, might want to take a look at this," a senior enlisted operator called to the U.S. Northern Command director of operations. Air Force Major General Donna Zurich, wearing a tailored green flight suit and highly polished boots, moved to the operator's station. A shift supervisor was already there, looking at a display over the sergeant's shoulder.

"What's up?" Zurich asked, scanning the text and a maplike graphic of Southern California. A frozen, red-shaded icon was circled, indicating an aircraft of interest.

"Los Angeles Center's lost an airliner, it appears," the operator said. "Still cross-checking, but they think it might have gone down there." He pointed to a spot in the San Gorgonio Wilderness. A C-17 pilot earlier in her career, Zurich had flown over that area dozens of times and was familiar with the rugged, steep terrain depicted by dark colors on a tactical-chart flight display.

"Type of aircraft?"

"Boeing 757. Arcadia Airlines' Flight Six-Two-Three," the operator replied, pointing to a data block in the screen's upper right corner. "It was descending for landing at Ontario; in the weather and at night, of course."

Zurich studied the screen, then straightened and crossed her arms, thinking aloud. "Going to Ontario . . . in the gunk, at night . . . Why was he that far north?" she asked rhetorically. Nobody answered. She *was* a pilot, after all. They weren't.

"What do you think, guys? Any indications of a terrorist connection here?" she clipped.

"No, ma'am. The FAA's handling it as an accident, so far, and the NTSB's launching a 'go team' to the crash site." Any domestic airliner accident triggered an invesigation by the National Transportation Safety Board, which had teams of experts on standby, ready to deploy on a moment's notice. Soon, Zurich knew, that team would be joined by engineers and flight operations specialists from Boeing, the aircraft manufacturer, and Arcadia Airlines. It could be days before they'd release a preliminary probable-cause report . . . unless traces of explosive were found at the crash site.

She nodded, studying the screen again. "Sure looks like an unfortunate accident, but there could be other factors, too. Stay in touch with the FAA and NTSB. If they smell anything that looks like a bomb onboard or anything else out of the ordinary, let me know. And offer the FAA NORTHCOM's assistance, if and when they want it. Oh, and let's get this incident up on one of the big screens. I want everybody to stay on top of this."

Zurich returned to her director's post, which overlooked two tiers of workstations facing a large, gently curving, U-shaped wall blanketed by flat-panel displays. She logged the incident on her computer and flagged it for inclusion in the commander's morning briefing. Doodling with a pen, she jotted down bits of data the FAA had forwarded, then studied the information.

Two experienced pilots on the Arcadia flight. Both frequently flew that route. No radio calls prior to dropping off the controller's screen. Something in her pilot's gut didn't feel right. *Why would an airplane be several miles off-course, descending into known high terrain!* Either two good pilots were flying heads-up-and-locked, not paying attention to where they were . . . *Unlikely. Damned unlikely, when descending into the Banning Pass!*

Zurich had flown that same route, preparing to land at the old March Air Force Base, and knew how dicey the route was. She checked a digital clock mounted above the big wall screens and mentally debated whether she should call the USAF Space Command ops center. In a few hours, the first-shift space guys would be arriving at the "Space Mall," a sprawling, aluminum-accented headquarters building next door. She decided to wait and ask one of her space cadet colleagues about the latest condition of America's GPS navigation satellites. Since that absolutely mindless North Korean nuke fiasco, GPS had been flaky, at best. She wondered if that might have something to do with the Arcadia accident.

APARTMENT OVERLOOKING LONG BEACH HARBOR

Roughly 50 miles to the southwest, Khalid's Hezbollah brothers were unleashing yet another horror. Three men worked silently, each activating a separate system connected to individual notebook computers. Ahmed, a fourth member and team leader of what the FBI called a Hezbollah "sleeper cell," watched anxiously. The oldest, he was tense and excited, but appeared calm and confident. By all indications, his team's sophisticated equipment was working its three-pronged magic. The first position was "spoofing" GPS signals in a relatively narrow beam. Tests the team had conducted from a rented boat weeks earlier had confirmed the spoofing beam, when fired across the busy Long Beach deep-water port, reached all the way to the horizon, a good thirty-five miles away, thanks to their apartment's twentieth-floor elevation above the port's water.

The other two "magic" boxes were now transmitting powerful beams of energy that jammed narrow segments of the electromagnetic spectrum, bands dedicated to marine radar and certain ship-to-shore satellite communications channels. For a long few minutes, any ship within forty miles or so of the Long Beach port would fall victim to false GPS navigation data, and both radar and communication systems jamming. In the darkness of this early morning hour, exacerbated by thick fog, those ships were electronically blind, mute, and, to some degree, lost.

"Ahmed, we have a likely target," one of the men announced. The team leader stepped to the GPS-spoofing console. Its operator simply pointed to an icon on a ship-tracking display. A commercial Web site collected and displayed GPS-transmitted position information from ships cruising up and down the U.S. and Canadian west coast, enabling maritime enthusiasts to track myriad cargo ships, oil tankers, container vessels, and large pleasure yachts.

Ahmed studied an icon for a few seconds, before summarizing. "Hmmm . . . thirty-three nautical miles offshore. On course for Long Beach, making about twenty knots, correct?"

"Right," the operator nodded. He slewed a pointer over the ship's icon and double-clicked the mouse. The Web-window refreshed, now showing a long list of data about that ship. "It's the *Dragon Wei*, a China-flagged container ship inbound from Shanghai. She's fairly new, very big, has powerful engines and is probably fully loaded, based on its open-ocean speed."

Ahmed nodded, smiling. Perfect! "Yes! Reel him in." The operator tapped a few keys and hit "Enter," triggering a sophisticated algorithm that slowly, inexorably altered the GPS "corrections" being received by the *Dragon Wei*. The digital streams were constructed to gradually alter the ship's perceived position. In less than an hour, the ship would be a few miles from the mouth of the Long Beach port, but its crew would see only a now-spoofed GPS position readout, which would lead them to believe the ship was still thirty-plus miles from shore.

The two men remained silent, each closely watching the ship's speed readout on the computer screen. If the speed decreased, then their jihad mission would fail. But the speed did *not* decrease. The *Dragon Wei* continued to steam at a solid twenty knots, even as it closed within ten miles of the port's entrance.

DRAGON WEI BRIDGE

"Why did you slow down?" the ship's captain, Tang Lee, barked in rapid-fire Mandarin. "We are still thirty-one miles from port!"

The *Dragon Wei* pilot rechecked his array of glowing, liquid-crystal color displays, then the position of his throttles. "Sir, we're at the proper power setting—no change for some time—but it appears we've lost headway. Our speed log does not correlate. It shows we were doing eighteen knots over the ground just a few minutes ago. Now we are slowed considerably. The current cannot be responsible, and the wind should be no factor," he responded anxiously.

The short, rotund captain seemed to ignore the man, but mentally reviewed his pilot's reply. He'd been into Long Beach at least twenty times and had never encountered wind or current conditions strong enough to slow the huge container ship this much. Something wasn't right. He studied the watch on his left arm, mentally calculating. They should be much closer to Long Beach by now.

"Radar! How far from shore do you show?" Captain Lee demanded.

"Sir, I'm showing nothing but a few ships to port and starboard. Nothing within thirty miles directly ahead," the junior radar officer said.

The bridge grew silent, waiting for the gravel-voiced captain's assessment. The six men stood at their stations, each monitoring a different set of displays. The ship's bridge epitomized modern seagoing vessels, equipped with the latest in radar, navigation, communication, and depth-sensing gear. And its crew comprised the best in Dragon Line's stable. All but the radar operator were experienced seamen, and each was a professional, the owner of specific skills honed by recent training.

"Communications! See if Long Beach has a position on us," the captain ordered.

"Aye, sir," the officer responded, then transmitted on the appropriate frequency. He waited, tapped a few keys, and transmitted again.

"Sir, there is no acknowledgment from Long Beach. Both of our satcom channels are filled with noise. I do not understand it . . ."

the comm officer finally responded, exasperated. He flipped a switch, allowing the hiss of dead-radio air to fill the bridge's speakers, as if to underscore his report.

Again, the bridge grew silent, except for the distant rumble of powerful engines. The captain grunted, then stalked to the navigator's position. The nav officer deferred, stepping to the side. His captain scanned the navigation screen, pivoted on a heel, and stepped to the radar station and again perused a flat-screen display. He paused and stared forward through the ship's expansive windows, hands clasped behind him. Ink-black darkness stared back.

"Forward lights!" he commanded. The electrical systems officer wrinkled a brow in question and glanced at the pilot, who shrugged. A series of computer commands turned the outside into a brilliantly lit fairway. But the powerful light beams were reflected back by dense, swirling fog. The pilot could barely see stacks of truck-size cargo containers on the foredeck.

"Lights off," the captain growled, turning away. He paced twice across the large bridge, stopped, and again consulted a large wrist-mounted chronograph.

"Scheduled port-arrival time is oh-three-forty," he declared. "The *Dragon Wei* has never, *never* been late! Set power to maintain twenty knots until we near the ten-mile barrier, then slow to port-entry speed," he barked, louder than usual. The other crewmembers exchanged furtive glances and turned back to their stations. Palpable worry hung in the quiet, conditioned air of the bridge.

The navigation officer tapped his keyboard furiously, trying to understand the GPS display's slow crawl. The system's position-and-speed readouts did not correlate with those of the backup inertial navigation system, which showed the ship was already inside that ten-mile barrier. But he dared not challenge the captain's orders, particularly after the stern officer apparently had perused both the INS and GPS readouts himself! He had been soundly chastised by Captain Lee days earlier, when he, the navigator, had questioned the reliability of the ship's GPS system. The short-tempered captain had threatened to replace the navigator at the next port, if the latter failed to "demonstrate a professional demeanor" on the bridge. That next port was Long Beach. This captain and the *Dragon Wei* were one, the navigator knew from experience. Criticism of the ship or its equipment was the same as slapping the captain's face. The navigator chose to remain silent, despite a growing fear that something was terribly wrong.

The *Dragon Wei*, indeed, was inside that critical ten-mile limit, still steaming at a breakneck twenty-plus knots instead of the legal three-knot port-entry speed. And its course was offset a few critical degrees to the left. The massive container ship was not aimed at the port's entrance and its intended berth, as its crew believed.

But a strict, hierarchical Chinese culture precluded further challenges to Captain Tang Lee's decisions, setting the container ship on course to assured tragedy. For his part, the captain dared not lose face by entering the port late, an inexcusable blemish on his faultless company record. Lee could not fathom the shame of being late. The *Dragon Wei*, at least under his command, would *never* be late for a port call.

LONG BEACH PORT CONTROL CENTER

"*Dragon Wei*, Long Beach Port! Acknowledge immediately!" No response.

"*Dragon Wei*, reverse engines *immediately*! Long Beach shows you at twenty-one knots inside the ten-mile arc! Slow to three knots *immediately*!" the harbormaster screamed into a radio microphone. For fifteen long minutes, the duty port controller had failed to raise the *Wei* by radio, prior to sounding an alarm, which roused the night-shift harbormaster from his office. The latter had seized the radio's microphone the second he'd digested the unfolding disaster. But there was nothing any of the port professionals could do but watch in horror as thousands of tons steamed directly toward the port's crowded docks.

Neither Long Beach Port's night crew nor the *Dragon Wei*'s bridge crew had time to untangle the high-tech electronic web Ahmed and his fellow jihadists had woven. Those three compact boxes, each controlled by an off-the-shelf computer and linked by a multiconductor cable to an innocuous antenna mounted on the roof of their apartment building, were in complete control of the *Wei*'s immediate future. With the ship's radar and communication channels jammed, the crew had no way to detect a fast-approaching shoreline, or to contact the port's controllers. And the GPS-spoofing system beamed false local position-correction messages to the *Wei*'s GPS navigation system. The ship's nav system showed the *Dragon Wei* miles at sea, when, in fact, she was within a few miles of Long Beach's crowded port.

An early, dense "May Gray" fog completed the scenario, preventing the *Wei*'s lookouts from seeing the jillions of Southern California lights along the coast. The ship, its crew, and $129 million of cargo were completely dependent on falsified GPS signals, left without the backup and assistance of radar information and satellite communications links.

At 0306 Pacific Daylight Time, the Chinese merchant vessel *Dragon Wei* brushed the end of a manmade sea wall and plowed into a newly built, offshore liquid natural gas-loading terminal, still making twenty-one knots. Due to the thick fog, not a person on the ship had seen the complex's huge storage tanks and complex pipe network prior to the container ship's impact. Kinetic energy imparted by the ship's combined speed and mass destroyed multiple pier supports, rupturing twelve-inch-diameter pipes carrying ultra-cold liquid natural gas. The LNG instantly vaporized and mixed with the air's oxygen, creating a highly volatile cloud. Failed steel supports screeched at earsplitting volume, mixing with the thunder of shipping containers torn from their tie-downs and tumbling into the terminal's densely packed infrastructure.

On the *Wei*'s bridge, a few shouts and a cacophony of automatically activated proximity alarms warned of the ship's impending impact, but nothing could be done in the few seconds that intervened. Captain Lee screamed *"Emergency Stop!"* But before they could respond, all seven men on the expansive bridge were hurled through thick Plexiglas windows as the ship decelerated from cruise speed to zero knots in a few horrible seconds. The vessel's nose reared up as it ran aground, then tilted to starboard, pitching bodies into the terminal's screaming, tortured steel.

Crewmen who survived the initial impact were consumed by a

nuclear bomb-like explosion triggered the instant a spark hit the deadly LNG-air cloud. A fireball engulfed the ship and LNG complex, expanding at lightning speed to rip into domed tanks, then nearby squat cylindrical tanks storing millions of gallons of crude oil, jet fuel, and gasoline. Tangled, broken pipes spewed natural gas into the conflagration, triggering a firestorm and incessant, rapid-fire explosions that swept inland. Windows shattered for miles around, from the hills of Palos Verde to high-rise condominiums and hotels in nearby Long Beach. A tsunami of demonic, black-laced fire raced up highways and streets, engulfing hotels, office buildings, retail shops, landscaped medians, and hundreds of vehicles that routinely served one of the nation's most critical seaports.

Fully one third of America's goods passed through Long Beach, which was now a hell of fire, serial explosions, and deadly flying missiles. Driven by the blast's massive shock wave, cars and trucks flew through the air, carrying their occupants hundreds of feet before slamming into freeway overpass pillars, other vehicles, and buildings. Soon, the rising, billowing cloud of fire rained flaming debris into fuel dumps, massive oil-storage tanks, and pipe complexes a mile or more to the north. Secondary explosions in these areas decimated structures well inland and threatened to tear into the nearby Port of Los Angeles.

APARTMENT OVERLOOKING LONG BEACH PORT

The staggering series of detonations far exceeded anything Ahmed or his team could have predicted or dreamed. Before the *Dragon Wei* slammed into the LNG terminal, all four of the jihadists had rushed to tall windows overlooking the busy port, anxious to witness the Master's brilliant plan come to a fiery, deadly conclusion. For long seconds, they found their view obstructed by a thick, roiling fog bank that blanketed the coastline. But an instant later, they witnessed far more than they'd anticipated.

When the terminal's LNG pipe complex and spherical storage tanks exploded, creating a billowing mushroom cloud of fire and smoke and steel, the men had less than two seconds to squint through the fog, hoping to marvel at the awesome display they'd unleashed. But in a heartbeat, the firestorm swept past them, destined to consume most of Long Beach and the surrounding areas. Before Ahmed and his team could shout for joy, an explosion's powerful shock wave arrived, shattering the apartment's windows.

Millions of tiny shards filled the room at sonic speed, shredding the flesh of four jihadists. Ahmed died when a three-inch dagger of thick glass sliced through his right eye and out the back of his skull, spewing a cloud of red. The blast's concussion threw his body across the room, slamming it through an interior wall. Blood and gray brain matter surrounded the hole his body created. His legs jerked and twitched, then fell still as glass and debris rained on them.

There was no shock at the sight of Ahmed's twisted body hanging from the hole, though. Two other Soldiers of Allah had met similar fates. Only the GPS-spoofer operator survived that initial high-energy blast, thanks to his randomly chosen position near the window's metal frame. His body had been somewhat shielded by thick, drawn-back drapes, but only partially. In shock, his eardrums shattered and a once-handsome face hanging in bloody strips, the young man staggered a few steps, then slid to the floor, back against an overturned sofa. His world was silent; he could hear nothing. He numbly lifted a hand to his face, trying to make sense of a hole where his nose had been seconds before. Through a haze of crimson, he watched his life stream from hundreds of cuts on his chest, stomach, and legs. They were bare, stripped of the T-shirt and jeans he had worn. He was overcome with an uncontrollable desire to sleep.

As his world faded from gray to black, the young jihadist had the vague, yet blissful impression that he surely must be dying. He tried to smile at a faraway, amusing thought that he must look like hell. He rejoiced, though, knowing that he was now a heroic *muja-hedin*.

6 SHOOTOUT

WHITE HOUSE LIVING QUARTERS

"Oh, God, no!" President Pierce Boyer whispered into the phone. "Is that confirmed, Vic? Could it be some kind of mistake . . . ?" His voice trailed off, trying to make sense of what he'd just heard. He was still trying to wake up.

"I'm sorry, Mr. President. It's definitely confirmed. It's already all over CNN and Fox," replied Victor Casebeer, America's Secretary of Homeland Security. A former Justice Department chief, Casebeer had recently been appointed to head the Department of Homeland Security, a job he'd reluctantly agreed to take, thanks to the president's "strong request." This morning, he desperately wished he'd stuck with his first answer: "No, thanks, Mr. President." He was staring at a muted TV screen filled with jittery images of a horrific firestorm. Firefighters silhouetted by roaring flames ran back and forth, many dragging heavy hoses. It was still dark in California, he noted. And very foggy.

Boyer ran a hand through unruly, sleep-matted hair, trying to grasp the nightmare he had awakened to only minutes ago. He drew a bone-weary breath, struggling to control a torrent of wild thoughts doing laps around his brain. "Good Lord! Long Beach Port devastated, hundreds, maybe thousands dead and injured. . . . And an airliner down around the same time?"

"Correct, sir. We don't know if the two are tied together, but the Long Beach incident and airliner crash happened within minutes of each other. Apparently, both the ship and the airplane were off-course, although that could be strictly coincidence," Casebeer cautioned.

"These days, I don't believe in coincidences," Boyer replied forcefully. "Okay, we have to get a handle on this mess ASAP. Pull together everything you have and get over here by seven-thirty, prepared to brief the security staff. And pass the word: We consider both of those 'accidents' acts of terrorism, until proven otherwise. Got it?" He added a few administrative details, then signed off.

Boyer immediately punched a speed-dial number on the phone beside his bed. Absently, he noticed his wife had rolled over and was eyeing him with concern. A wisp of light-brown curl fell across her forehead.

"Gil! You've heard . . . ?" Boyer barked into the phone. "Of course . . ." he added a second later. His chief of staff, Gil Vega, had obviously been awake for some time, working the issue. "Listen, notify the security team that we'll convene in the Situation Room at seven-thirty. I want whatever everybody can pull together by then. Forget the damned analysis; raw information is better than nothing. We've *got* to get something out before eight, or the press corps'll be tearing the gates down. We're on a short fuse. . . . Yes, definitely. Set up a press briefing for nine, and tell our people that nobody, *nobody* says a damned thing before then! No stupid 'leaks' that're nothing but speculation, okay? . . . Yeah, good point. Start the ball rolling to get me fifteen minutes of TV time, and work with Mandy to get the basics of a quick speech together. . . . Right. Just a few facts to give 'Joe Six-Pack' a warm fuzzy, until we figure out what the hell's happened. We'll fill in the blanks after the seven-thirty briefings."

Boyer plopped onto the bed's edge, shoulders slumped. The First Lady sat up, leaning against the headboard, and put a hand on the president's shoulder. "More bad news?" she asked quietly.

" 'Fraid so, hon. A Chinese container ship plowed into the oil and gas facilities in Long Beach Harbor, causing one hell of an explosion. Lot of damage, people killed. . . . Christamighty!" Boyer shook his wild hair, twisting to face his wife. She searched his eyes, seeing nothing but deep pain. The man was aging rapidly, she noted, alarmed.

"And a damned airliner crashed into a mountain near Ontario. Two 'accidents,' within minutes of each other, both in Southern California . . . Doesn't look good."

His wife kissed her husband's cheek, then rolled over and lifted the house phone on her side of the bed. "Good morning, Trish. I know it's early, dear, but could we get some coffee? Better make it strong. . . . Yes, I have . . . I don't know, but you could be right. This might be nine-eleven all over again."

STRATCOM COMMANDER'S OFFICE

Scrolling down the flat-panel computer screen, General Howard Aster skimmed a preliminary report from an Air Force Space Command field team deployed to Southern California, mentally logging specific phrases: . . . *Boeing 757 off-course approximately two miles;* . . . *normal angle-of-descent; no deviation of flight path prior to impact in San Gorgonio Wilderness. . . . No in-flight problems reported by crew. No radio transmissions between crew and FAA controllers immediately prior to crash. . . . Container ship Dragon Wei entered Port at high speed. Crew did not respond to radio calls . . . Dense fog and thick cloud layers blanketed large areas of So. Cal. at the time of both incidents. . . .*

A toss-off sentence in the final paragraph caught his attention, prompting a reread: *Preliminary law enforcement investigations suggest that both incidents may have a terrorism connection. In support of that assessment, we are pursuing a theory that persons unknown may have capitalized on the currently degraded GPS satellite-based navigation network, which may be adversely influencing a variety of guidance systems. This line of investigation is very preliminary, unconfirmed, and considered classified . . .*

Shortly after details of the Arcadia Airlines 757 crash and the container-ship disaster at Long Beach's busy port hit the news, Air Force General Erik "Buzz" Sawyer, the blue-suit Space Command chief, advised STRATCOM that he had immediately deployed a four-person team to southern California.

In recent years, provision of Air Force, Army, and Navy space experts had become standard operating procedure during large-scale domestic-security operations, ranging from natural disasters to terrorist strikes. Joint-service space teams provided a variety of services, from satellite phones for on-scene commanders, to timely weather data downloaded directly from meteorological satellites.

The four-star commander of STRATCOM stood and stretched, then started pacing back and forth across a spacious office. Both hands were jammed into the pockets of his dark-blue uniform pants. He finally stopped, picked up a coffee cup, and turned to a large window behind his desk. Sipping the hot brew, he squinted at the sun-tinged Nebraska countryside for several minutes. Early morning sunshine was already burning the dew from Offutt Air Force Base's spring-green grass, he observed absently.

"Burner, I've got a bad feeling," Aster finally said over his shoulder, still staring through the window.

"Sir?" the muscular general's aide responded, brow furrowed. Lieutenant Colonel "Burner" Burns glanced toward the STRATCOM chief. Burns had been sorting documents for Aster's signature, spreading them neatly along one edge of a huge oak conference table. Even a year-old, critical "space war" hadn't slowed the paper-generating bureaucracy of STRATCOM headquarters. Getting the boss cornered long enough to scrawl his signature on a never-ending stream of memos was a continuing challenge for Burns, a former B-2 bomber pilot and squadron commander.

Being an aide and "chief gofer" to STRATCOM's commander hadn't ranked high on Burns's want-to-do list, but he'd reluctantly accepted the joint-service assignment. Flying a stealthy, long-range bomber suited him far better.

Burns's reluctance to leave a stealth-bomber flying position was salved somewhat by the understanding that his real STRATCOM job actually *was* more important than pushing paper for the general. He shadowed the commander everywhere, carrying awesome go-to-war codes in an aluminum briefcase popularly known as "the football." Should global conflict begin in earnest, both the STRAT-COM chief and the President of the United States would turn to these "shadows." Burns and his counterpart in Washington would break out the "go-codes" that could unleash America's nuclear might against a decidedly unlucky enemy.

"That Space Command report from the team in California . . . Did you catch the reference to 'degraded GPS infrastructure'?" Aster asked, turning to the chocolate-skinned, well-built aide.

"Yessir, I did."

"What do you make of it?"

Burns carefully set aside the sheaf of documents, hesitating. "Sir, if I were a betting man, I'd say both of those accidents in So-Cal were tied to bad GPS data. Don't know how, but . . ."

"Rog! Me, too," Aster interrupted. "Call it an old fighter pilot's gut instinct, but I don't think those disasters were coincidences. Too many common factors, right down to the crappy weather, the timing, no reported problems from either the airplane or the ship— no hint whatever of something wrong." The former Air Force Academy basketball star resumed his pacing, still carrying the stained coffee cup.

"We know that damned North Korean nuke charged up the ionosphere, creating a shield that's hard for electromagnetic energy to penetrate. That, in turn, is attenuating GPS signals to varying degrees all over the country, causing position and timing data dropouts. The FAA administrator was so antsy about the current sorry condition of GPS data that she seriously considered grounding every civilian airplane in the country for safety reasons. At least until GPS got healthy again."

Burns knew all this, but the boss was obviously thinking through the myriad ramifications of a degraded GPS network, resorting to his now-familiar, step-by-step method of out-loud reasoning. The general often did his deep thinking while pacing.

"The airlines screamed and ran to Congress," Aster continued, one hand jammed in a pocket. "Most of those carriers—the big ones, anyway—were finally eking out a profit, after years of bleeding red ink following the nine-eleven attacks. Of *course* they insisted on staying in the air! Only damn way they could make any money, so the FAA caved and let 'em keep flying. In retrospect, I'm not so sure that was the right decision. What's your take, Burner?"

Aster respected Burns's level-headed reasoning—even though his muscular, highly intelligent executive officer and aide *was* a bomber puke. Aster had always flown fighters. While the two combat-flying communities rarely viewed the world the same way, they each had an unspoken respect for the other's capabilities and contributions. True, it was unusual for a four-star general to bounce ideas off a lieutenant-colonel aide, but Aster was a rare commodity in the twenty-first-century Air Force, a true-blue leader, not the more-typical "manager." He was committed to growing the service's next generation of leaders, and that translated to mentoring smart young officers, like Burns. Besides, the general always came away from their brainstorming interchanges with a clearer picture of a knotty situation. He was convinced Burns would sport stars on his shoulders in the near future.

"Yes, sir," the stocky black man nodded confidently, his shaved-bald head reflecting the overhead fluorescent lights. "I believe it *was* the correct decision. Any airliner certified in this country is equipped with an INS [inertial navigation system] that's coupled to an integral GPS receiver. And since the FAA started shutting down ground-based navaids a few years ago, every pilot out there's been relying on GPS. But he also still has an inertial unit in his airplane, cross-checking every GPS nav solution. As I see it, sir, the INS bridges gaps in GPS data anyway, so no problem. Airplanes still get where they're going, even if they do suffer GPS dropouts."

"Hmmmph," Aster grunted. He stopped pacing and downed the remaining coffee, letting a long silence develop, before turning to Burns. "Can't argue with your logic, Burner. And yet, something about this degraded GPS situation and those two accidents in California is bugging the hell out of me. Can't put my finger on it, though.

"Look, I admit; I'm worried," he added, waving the empty cup. "That business of 'sliming' the Chinese nanosats still concerns me big-time, but it's a done deal. Hell, we may have committed an act of war and don't know it! We're doing stuff that may have huge un-intended consequences 'cause we're literally flying blind here! We don't know *what* we don't know.

"Yeah, we'll figure out what happened in California eventually, and those disasters may not have had a damned thing to do with a degraded-GPS posture. Maybe I'm just being paranoid," he contin-ued. "But STRATCOM is in dire need of some sharp wizards to help us sort out this top-level space stuff. Not the technical ones-and-zeros aspects. We already have great tech experts down there in the wargaming center and out in California, working those two in-cidents from the geek angles.

"Buzz Sawyer's space people have done a hell of a job helping our gamers and ops folks noodle through those space-techie issues, too. No problem there. But *I* need big-brain help. Somebody who can help *me* sort through top-level space-policy and-doctrine issues, things we've never faced before. And maybe a few specialists to help our 'gamer Red teams work out how an adversary might capi-talize on this god-awful 'degraded space infrastructure' mess that just seems to get worse by the day!" Aster was practically spitting, exposing a nagging frustration.

Jabbing his empty cup toward Burns, the general abruptly snapped, "Find me those big-picture space experts, Burner. And get 'em here ASAP. Copy?" Not waiting for an answer, the general headed for his office door. A refill of dark-roast Colombian java had assumed top priority.

Within the next twenty minutes, Burns found Aster's experts. Two of them would soon be headed for the Colorado Springs air-port, en route to Omaha. He'd called the National Security Space Institute in the Springs and spoken to the NSSI's full-colonel commandant, an officer he knew from past association. Burns had attended the NSSI's two-week Space Operations Course as a B-2-flying major a few years earlier. The course not only provided a con-cise warfighter's understanding of space-asset capabilities and limitations, but also plugged students into an expanding personal network of space expertise that would be tapped throughout their careers.

Burns reviewed his notes at a small desk tucked in one corner of a well-appointed foyer outside the STRATCOM commander's office. He wore an olive-drab flight suit stretched tightly across a muscular torso. Black oak leafs were sewn atop each shoulder of his one-piece "fly bag," and the right upper arm sported a bomb-in-bullseye patch. That shield-shaped patch subtly announced he was a graduate of the coveted Weapons School at Nellis Air Force Base, near Las Vegas, Nevada. On the flight suit's chest, above an angled zipper pocket, an embroidered name tag included his "Burner" call sign. The stocky Afro-American officer had a reputation for being intensely focused and perpetually in a hurry. That full-throttle intensity and rapid stride had earned him the "Burner"—as in "afterburner"—moniker early in his Air Force career.

Notes in hand, he waited until the door to Aster's private lavatory opened, accompanied by the unmistakable sound of plumbing at work. He raised an eyebrow to Annie, the general's executive assistant, seated at her desk near Aster's open office door. The pert, immaculately dressed, fifty-something woman smiled, nodded crisply, and turned back to her computer. Burns entered the commander's office in time to intercept Aster.

"Sir, a couple of NSSI policy and doctrine experts will be up here tomorrow," Burns announced.

"NSSI? Help me out, Burner. You're dealing with a fighter pilot, so keep it simple," Aster growled, long steps taking him to a broad desk.

"Sorry, sir. The National Security Space Institute in Colorado Springs. It's the 'space schoolhouse.' It provides professional leadership education for the entire national-security space community."

"Oh yeah, sure," the general nodded, sliding into a plush, wheeled leather chair. "The Secretary's been on my butt to get down there and partake of its space ops executive course." He shot Burns a lopsided smile, adding, "Probably should have done that before we got into *this* shit-storm, huh?"

Burns shrugged and grinned. "Well, the course sure helped this ol' bomber pilot understand space mumbo jumbo a little better. There's a slim chance a fighter pilot *might* get something out of it, sir." Aster flipped a thumbs-up touché and chuckled.

"Sir, NSSI's commandant is detailing two space-smart officers to help our wargamers," Burns continued. "One's a former 'Space Aggressor' squadron commander, who knows more about satellite system vulnerabilities than the bad guys do. And the other's a top-notch space operations expert."

"How about a space doctrine guy? Damn it, I need somebody to help the ol' STRATCOM commander, remember?" Aster barked.

"Uhhh . . . sir, that's our *third* NSSI person. And your 'guy' is NSSI's female chancellor, Major General Erikson. She definitely has a good handle on space law, policy, doctrine, and historical precedents, and is certainly the institute's best 'think tank' sort." Burns hesitated, caught momentarily by Aster's sharp glance and strange half-smile. "Sir . . . ?"

"Nothing," Aster flicked a hand and turned toward the expansive window behind him. "Just surprised to hear Viking's the NSSI chancellor. I've worked with her before. She was my intel ops chief when I was a wing commander during Gulf War II. Sharp kid." Aster absently rapped four fingers on the desk top, staring out the office's big window. "Dawn Erikson. I'll be damned . . . ," he said softly.

Burns nodded, mentally logging the boss's odd reaction and tone. It seemed incongruous with the four-star's usual no-nonsense demeanor. He brushed aside a silly brain-flash.

"Sir, General Erikson's at Los Angeles Air Force Base today, wrapping up meetings at the Space and Missile Systems Center. Her aide said they'd catch a United Airlines hop up here late this afternoon."

"Good! I sure hope Viking's got her shit together on this space business," Aster mused. "She was a very smart intel officer, but I had no idea she'd wound up as a space cadet."

"Sir, space and intel are as tight as fuzz on a peach these days," Burns said, cracking a rare smile. "General Erikson gave our NSSI class a superb lecture about the space-intel link. If that was any indication, I'd say she definitely has her 'space stuff' together."

"Thanks, Burner. When the NSSI folks get here, get the two tech-smart guys into the wargame downstairs, and give me a heads-up. I want to see General Erikson ASAP," Aster clipped, slapping his desktop.

TERMINAL SIX/LOS ANGELES INTERNATIONAL AIRPORT

Army Major Hans Richter opened a rear car door and took the thin leather briefcase from his boss, Major General Dawn Erikson. A slim, blond woman slid from the seat, still talking rapidly into a palm-size multifunction communicator she held to her left ear. At the open trunk, Richter accepted two bulging "wheely" suitcases from the driver, shook hands with Airman First Class Donley and thanked him for the ride. Standing a smidgen over six feet tall and tipping a rock-solid 210 pounds, the Ranger-qualified officer didn't bother dragging the luggage. He tucked one bag under an arm and grabbed the other's handle, then ushered Erikson toward a sliding glass door.

Inside, Richter groaned. Instead of the usual single lane of meandering switchbacks marked by waist-high poles and stretched fabric belts, United Airlines had reverted to a series of parallel lines in front of about nine check-in counters. Every line was jammed with at least a dozen passengers waiting to check luggage and get their boarding passes.

"Hans!" The major turned to see his boss waving two slips of paper, still talking to her communicator.

He retrieved the slips, silently mouthed, "All right!" and gave her a partially raised fist in thanks. Some thoughtful airman at L.A. Air Force Base had printed out Erikson's and his boarding passes, checking in via United's Web site, before they'd left for the airport. He headed for the "Online Check-In" counters, both devoid of passengers. Erikson followed, still talking into her phone and digging in a purse for a military identification "smartcard."

Minutes later, minus two pieces of luggage, Richter and Erikson were walking slowly toward the terminal's security-screening area, passing a bank of floor-to-ceiling windows. The sweeping expanse of glass overlooked a four-lane airport access road jammed with noisy cars and buses. Even inside, they could hear the traffic and smell L.A.'s smog-choked air. Richter carried both his own soft-sided briefcase, which housed a thin notebook computer and reams of documents his boss seemed to collect, as well as the general's petite, expensive leather case.

Erikson cursed softly, shook her head, and stepped over to one of those windows. Today's wireless communicators were far superior to the old cellular telephones of the mid-2000s, but signal dropouts inside steel-and-concrete buildings still occurred with aggravating regularity. Richter sighed, clamped his and her briefcases between his ankles and crossed his arms. He stood a few feet behind Erikson while she redialed and continued the conversation, now facing a window to ensure a solid wireless connection.

Staring absently through the plate glass, Richter reflected on his current life's situation. He still marveled that the fickle military assignment system had actually worked in his favor this time. After three quick-turn combat tours in Afghanistan and Iraq, the muscular Ranger had faced a tougher adversary than any al Qaeda insurgent or Taliban fighter. His petite, lovely wife had given him a heart-stopping ultimatum: Either get a stateside assignment having some semblance of stability, so he could be a *real* father to their three children, or resign his commission and get a civilian job. Otherwise, she was filing for divorce. Period.

She might as well have hit him with an infantry trenching tool. He'd been so busy leading troops in combat, ferreting out and killing bad guys, that he'd completely missed seeing how stressed and fed up his bride of fourteen years had become. She was at the end of her rope, and apparently had been for some time.

Richter was the quintessential warrior of his era. After graduating in the upper third of his class at the U.S. Military Academy at West Point, he'd gone directly to Infantry Officer Basic Course, then Ranger and airborne schools. Training was tough, sure, but he'd thrived on the challenge of slogging through stinking swamps in downpours, slipping quietly through dense, snowpacked forests and making night jumps from airplanes, carrying a fifty-pound rucksack and rifle. And leading troops in combat was about as good as life could get, in his judgment. But family took precedence over career, a commitment he'd made to Cindi the day she pinned on his gold second lieutenant's bars, hours before they exchanged vows in the West Point chapel.

With more than a little behind-the-scenes arm-twisting, he'd wrangled a posting to the Army's still-maturing space corps and moved his family to Colorado Springs. Within a month of arrival, the Army Space and Missile Command's local commander had offered Richter an unusual two-year "joint" position as the aide-de-camp for Major General Dawn Erikson, an Air Force Reserve officer heading the National Security Space Institute there at Peterson Air Force Base. He'd been reluctant to leave the familiar Army environment, but decided the job more than satisfied Cindi's adamant criterion for domestic stability. He moved to the NSSI's brand new building, and for the first time in many years, Richter was home by 6 p.m. on most nights.

As a reservist, General Erikson was only on-site at the NSSI facility about a week each month, which simplified Richter's job immensely. However, as the institute's chancellor, she traveled to Washington, D.C., frequently, meeting with political leaders and "spreading the gospel of professional space education," in her words. Richter accompanied her on these trips, when it made sense, but he often logged desk and instructor time at the NSSI schoolhouse, instead.

Richter glanced toward Erikson and grinned as she caught his eye, raising one finger. *Right! No way only one minute!* he thought. His boss spent most of her waking hours in meetings, talking on the phone or text messaging, and answering e-mail via that damned do-all communicator. He watched as she turned back to the window, still talking and waving one hand as she drove home a point to who-knows-who on the other end of that link. *Hell, she's gotta be fifty years old, and still good-lookin' enough to cause traffic pile-ups. And energy? The lady runs my ass into the ground!* he thought admiringly.

Indeed, Dawn Erikson was a regular on Indiana's 10-kilometer race circuit, managing to stay in excellent physical condition, despite a grueling business schedule. At a slim, ramrod-straight five-foot-seven, she presented a small waist, short, yet feminine hairstyle, and a youthful, attractive, angular face dominated by killer pale-blue eyes that screamed of an ultra-keen intellect. The summer-gold blond hair and ice-blue irises were products of Scandinavian lineage. In her Air Force blues, dual silver stars on each shoulder, *plus* Nordic good looks, "Viking" Erikson commanded undivided attention.

Today, even though Erikson was traveling in casual civilian clothes, she still drew admiring glances. She wore yellow slacks and a matching pullover knit top covered by a short-waisted jacket. An efficient-looking pear-shaped purse hung over her left shoulder.

Richter, too, was dressed in civvies, thanks to a Defense Department regulation that stated military officers would not travel in uniform on commercial flights. No sense in being attractive targets for terrorists or would-be hijackers. A few Islamists had attempted to kidnap military officers in various countries, but Richter wasn't aware of any incidents in the U.S. Nevertheless, he welcomed the Pentagon dictum, preferring to blend in with the traveling public. And when crammed into a too-small airline seat, civvies were a hell of a lot more comfortable than a uniform. To that end, he wore a ball cap with "Broncos" scrolled above the curved bill, a pair of Levi's stonewashed blue jeans, a long-sleeved denim shirt, and Nike running shoes.

Richter stifled a yawn, scooped up the two briefcases and stared absently at the bustle of traffic and passengers outside. Ten feet away, a late-model Toyota pulled up to the curb and disgorged four young men. From the trunk, they each retrieved a golf bag and headed for the terminal's automatic sliding door. Richter barely noticed them, at first. Then something clicked in his combat-honed mind, causing the thick-shouldered officer to turn and watch the four men. Without a word, they went to separate check-in lines in front of the busy United counter.

Richter's eyes wandered across the crowd of passengers, surreptitiously keeping the four copper-skinned, dark-haired men in his field of view. Dressed in white shirts, dark pants, and loafers, the four could be a team of some kind. They silently shuffled along in their respective lines, but kept glancing across at each other and over their shoulders. None smiled. One, standing in a longer line than the other three, met Richter's gaze, but quickly looked away.

An unexplainable something gnawed at the young major. Something not right. What, he couldn't say, but the hair on his neck bristled. He had learned to trust that feeling. Maybe it was an Army Ranger's fine-tuned sixth sense, or a momentary flashback to patrols on the dangerous streets of Baghdad, but something had twanged his heads-up antenna. *Why aren't those dudes all in the same damned line, if they're traveling . . . ?* Instantly, it hit him. *Golf bags!* Four golf bags, all exactly the same. *And no other luggage . . .*

Richter smoothly stepped to Erikson and took her right arm, above the elbow. Still talking, she threw him an irritated, unbelieving glance and jerked the arm. Richter held on firmly, and whispered through tight lips, "Ma'am, move! *Now*! No questions; just *move!*" He stepped around to her left side, between Erikson and the crowd of passengers, maintaining a grip on her right arm, his across her back. He forcibly nudged her away from the windows, causing her to almost stumble. He tightened his arm, holding her erect and moving forward.

The officers crossed in front of two young security guards standing beside a metal support post. Both had been talking quietly while continually scanning the crowd. Both wore dark-blue uniforms with pantlegs tucked into black, spit-shined combat boots. Heavy black belts sported radios, holstered 9-mm sidearms and snap-buttoned leather ammunition and handcuff pouches. Muscular arms stretched rolled-up, short-sleeved shirts. Sewn-on name tags and a shield-like badge adorned their shirts' chest. A leather sling looped over each guard's right shoulder was attached to a short-barrel M468 automatic rifle, Richter noticed.

One of the young guards winked at Erikson as she and Richter passed, prompting a subdued hoot of laughter from the other. The blond general missed the wink. She was still talking on the communicator, while shooting ice-eyed daggers at her aide, trying again to wrench her arm free of his grasp.

Richter also missed the general's withering glare. He had turned in time to see one of the swarthy young men flip the top of his golf bag open and yank out a Kalashnikov AK-47 automatic rifle. The would-be passenger dropped the golf bag, turning as he brought the rifle up to a waist-high firing position. Behind him, Richter heard a shout from one of the guards, then an ear-splitting *bratttt!* from the full-automatic rifle. Instantly, Richter dropped to the floor, both arms wrapped around Erikson, his back to the crowd of now-screaming passengers. Erikson tried to yell, but a massive arm had covered her head and mouth. *Braaaatttt!* A longer burst from another AK-47 sent one of the young guards sprawling across Richter and Erickson, eliciting a yelp from the lady-general. Richter jerked and grunted in pain, but tightened his bear hug on Erikson. He was lying on his left side, body and legs pinning the woman to the floor. He lay very still.

Major General Dawn Erikson was petrified, her face buried in Richter's chest. She could feel the big man's heart pounding and smell a hint of aftershave or men's cologne. Instinctively, she tried to push away from Richter's burly chest. The major's arms tightened around her, his shoulders hunkered forward and curled around the slim woman, shielding her from the chaos behind him. Then Erikson smelled the distinct, biting odor of spent ammunition, triggering a flash-memory of Iraq.

"*Shit!* I'm hit," Richter gritted in a whisper, more to himself than for Erikson's benefit.

"Hit? Where?" she whispered to his chest. He never answered, prompting Erikson to twist her head enough to see under the man's arm. *Oh God! Don't die, Hans!* she thought, a wave of panic washing over her. Richter's blue eyes, shaded by the ball cap pulled tight over a shaved-bald pate, were wide and bright, still sharp—and flicking back and forth. They were looking past her. *Thank you, God!* she prayed silently.

In what was left of the broken and cracked glass, a reflection showed the four shooters waving their weapons. They were screaming in Arabic-accented English, cajoling a herd of confused passengers to bunch together in a compact group. Richter froze, holding his breath as he saw a terrorist swing an AK-47 in his and Erikson's direction.

"Do *not* move, ma'am," he hissed. "*Freeze!*" He fervently hoped the growing spot of red beneath his throbbing calf would convince that shooter he was no threat.

The slim general stopped breathing, her eyes squeezed shut, trying to melt into the floor behind her aide's bulk. She fully expected to feel fire-hot bullets rip into Richter's and her bodies. Seconds passed as hours; neither officer dared breathe. They were partially covered by the airport guard, who had been thrown backward by the initial volley. He had slammed against the window, the lower portion of which was partially intact, despite multiple spider-webbed bullet holes.

Back arched across Richter's and Erikson's hips and legs, the guard was sprawled at unnatural angles, eyes locked open in a dis-believing, lifeless stare. His rifle had fallen to the floor and lay be-hind Erikson, inches away from Richter's hand.

The Army major's eyes were locked on the cracked window, con-tinually scanning the reflection. Barely audible, he whispered to Erikson. "There's an automatic rifle right behind you, ma'am. Hold it! Do *not* move! I'm going to pull it toward your back. When I tell you to—and not before—I want you to use your left hand to slip that nine-millimeter out of the guard's holster. *Don't move!*" Richter hissed as Erikson started to shift her head.

The aide took a slow, ragged breath, gritting his teeth and fight-ing through a wave of pain in his hip, where he'd slammed onto the tile floor. He was surprised the wound in his leg did *not* hurt. It was just numb. Carefully, he released the woman under his right arm and inched his fingers to the guard's rifle. His eyes never left the mirrored reflection.

Maybe thirty or forty-five seconds had passed since the first shots were fired. Behind Richter, the four terrorists were screaming at terrified passengers and ticket agents, trying to herd them into a cowering, tight group. An elderly man barked at one of the gun-men when the latter shoved a frail, gray-haired woman. The bold retort earned him a smack to the side of his head from an arcing rifle butt. The elderly man crumpled on the spot and didn't move, causing the woman to scream.

It was the distraction Richter needed; he had the guard's rifle in hand. He waited until all four terrorists were again occupied, then whispered, "*Now!* Grab that nine-millimeter!" Erikson raised her head slightly, stretching her left hand to smoothly slide the pistol from the guard's holster. Her heart was beating wildly, but her hand was steady, she noticed absently. Holding the cold, black pistol be-tween her and Richter was surprisingly comforting. A full clip was jammed into the handle.

"Now what?" she whispered. She had seen just enough to know horror was unfolding a few feet from where they lay, partially covered by the guard's unmoving body. Erikson's thin legs were stinging-numb from the combined weight of the guard's and Richter's heavy limbs pressing hers against the cold floor.

"Ma'am, listen to me. Those son-of-a-bitches are going to kill those people, then come over here and put a few slugs in us, just to be sure," Richter whispered, eyes still locked on the glass's reflection, tracking the gunmen's every move.

Erikson's lips moved, but only a squeak escaped. Her ice-blue eyes were huge, both pupils dilated. She forced her tousled hair closer to Richter's face. "So, what do we do?" she mouthed.

Richter gritted his teeth, fighting a stabbing pain in his leg. *Shit! It was numb a second ago!* he thought. He quickly explained their one-and-only option. "You'll have to move fast, or the bastards will get both of us. Got it?" Erikson's eyes searched his, apparently drawing strength from the solid man, who still cradled her tightly. Richter blanched and grimaced again.

"You're hurt, Hans!" He ignored her whisper, noting she was on the verge of panic.

"*Listen!* You've fired a nine-millimeter before. It has a full clip, so *carefully* charge it and flick the safety off." She complied, holding the weapon against his shirt to muffle the action's noise.

"Ready?" he asked softly, holding the general's eyes. A shock of blond hair had fallen across one of those wide baby blues.

She nodded. Richter whispered, *"Now!"* and swept the M468 automatic one-handed off the floor and across Erickson's head, half-twisting to aim, still prone. Simultaneously, she ducked under the arm, sat up and switched the 9-mm pistol to her right hand, planting the butt on her left palm in a smooth, continuous movement. She briefly steadied the front bead sight on the back of a gunman about eight feet to her left and squeezed deliberately. The weapon bucked as a small explosion smashed her ears. A second blast to her right caused her to flinch, but a microsecond later, she realized it was Richter firing.

Stark fear and her trademark Scandinavian determination combined to swing the pistol a few inches to the right, centering the front sight on a second gunman just as he pivoted, bringing the rifle's mean-looking muzzle up. She squeezed again, and immediately pulled the trigger a second time. The terrorist's hands flew skyward and he was lifted clear of the floor, flying into the knot of passengers that was diving amid deafening screams. His AK-47 clattered on the ceramic tile. Another short burst and she knew Richter had fired again, as well. Erikson swept her weapon back and forth, searching for a target, prepared to fire again. There was nobody standing. Still, her wide eyes scanned the area, waiting. . . .

She felt Richter move but kept her eyes on the crowd. The big man was lying prone, elbows spread to both sides, also sweeping his M468 across the knot of passengers. The white shirts of four terrorists were turning red. None of them moved. One was sprawled across several passengers, triggering shrill screams from a heavyset woman, who kept trying to push and kick the bleeding body off her. Slowly, Richter shifted the M468's muzzle to the right, away from the pile of screaming people.

Richter glanced over his shoulder at Erikson. "You okay?"

Erikson nodded. She knew damn well she couldn't make a sound. Her mouth was intensely dry, and her eyes kept snapping between Richter's steady, penetrating blues and the four terrorists' bodies, still afraid one of them would rear up and start shooting. She was holding her breath, conscious of a strong odor of gunpowder. And the 9-mm pistol kept moving, side-to-side, searching . . .

Richter half-rolled, freeing a hand, then sat up. He grimaced in pain, but slowly extended an arm and gently wrapped his fingers around Erikson's small right hand, which held the 9-mm in a firing-ready pose. His thumb flicked the safety closed.

"I've got it, ma'am. You can let go." The big Ranger's tone was quiet, barely audible above the shrill cacophony rising from the knot of passengers.

The general relaxed her grip, letting Richter take the pistol, then slumped backward, leaning on both hands. Her legs were still pinned, she noticed. A sharp ringing in her ears made it hard to hear Richter's voice. The constant ringing was an unpleasant counterpoint to an incessant, annoying scream that pervaded the terminal.

Both sounds seemed far away, though, as if in another world. She felt detached, watching the scene around her from afar. Everything seemed to move in slow motion.

Richter shifted, grimacing in pain as he untangled himself from the dead guard. That allowed Erikson to pull herself free of both men, as well. She tried to stand but couldn't; rubber legs refused to support her weight. On her knees, she turned to Richter, remembering that he was wounded and needed help.

But she was mind-numb, struggling to think, let alone help her aide. Not that it mattered. The Army major had already pulled his belt off and was using it to bind a folded red-and-white bandana over an oozing red smear on his right calf. *I didn't know men still carried handkerchiefs,* she thought, fighting a mental fog.

"Hans! You're . . . you're . . . shot!" she stammered. The voice was someone else's, a strange croak.

"Damn tootin', ma'am," he answered, teeth clamped. "And it hurts like hell! But, by God, we're alive!" He flashed her a toothy grin, then squeezed his eyes shut and swore as he cinched the belt tighter. Seated on the blood-streaked terminal floor with both legs stretched before him, Richter heaved a sigh, leaned back on one hand and gave Erikson an admiring look.

"Ma'am, you done good. Damned good," he said softly. She blinked, fighting very ungeneral-like tears.

Suddenly, Erikson and Richter were surrounded. The general recovered her faculties enough to collar an emergency medical technician, demanding he attend to Richter. Black-helmeted policemen were everywhere, it seemed, each bristling with armament and Kevlar vests. One apologized to Erikson and Richter as he stepped past them to gently cover the young guard with a plastic sheet. Airport officials and security forces appeared, trying to calm the agitated, terrorized passengers. Several people clutched cops' arms, babbling and pointing wide-eyed at Erikson and her aide. A TV crew pushed a large camera through the broken glass window, sweeping the crowd. The cameraman ignored the woman in yellow, a sheet-covered body, and an EMT attending to a man in a Broncos ball cap sitting on the floor. The camera was focused on the turmoil among passengers, cops, and airport personnel.

Dawn Erikson climbed into an ambulance with Richter, barking at an EMT who dared try to block her way. *Already back in command mode.* Richter smiled. He was very sleepy, lying on a clean-smelling gurney firmly anchored to the ambulance floor. His leg didn't hurt so much now. Whatever had been in the syringe that EMT had jammed into his arm was working nicely, he mused. That, plus the adrenaline drop after combat or other heart-pounding action made a man damned tired.

Major Hans Richter heard a siren and felt the vehicle moving, but it was tough to keep his eyes open. The last thing he remembered was General Erikson's pale-blues hovering above his own, so close hers were out of focus. Later that night, in the hospital, he would wonder if he'd only dreamed that the general had kissed him. Surely not . . .

7 PUPPETS AND MASTERS

WHITE HOUSE/NATIONAL SECURITY COUNCIL OFFICES

Attention locked on to a classified briefing document, Brigadier General Hank "Speed" Griffin absently reached for his handheld communicator, a combination crypto-secured cell phone, personal data assistant, and high-bandwidth datalink. The latest in high-tech, National Security Agency-issued communicators for top government officials, its flash-priority chime was sounding off.

"Air Force Liaison Office, General Griffin," he announced, eyes still sweeping the red-bordered file dominating his computer's screen. He could hardly believe what he was reading: a curt, impassive report about a North Korean nuclear weapon detonation high in the atmosphere. The blast's subsequent impact on both commercial and military electronics systems was surprisingly widespread, affecting a broad expanse of the northern Pacific region.

"Speed! Rattler here. Are you full-secure?" a voice replied, its tone urgent.

Griffin double-checked an icon on his handheld. "Yeah. I'm *Secure-Red*. What's up, guy?" Colonel Brian "Rattler" Rich was among the handful of USAF senior officers Griffin classed as one of the *really* good guys, a fellow fighter pilot from Speed's F-16 days. While

Rich had stayed in the fighter community, Griffin had diverted into the flight-testing world as a test pilot. However, they'd remained close friends and colleagues. Slightly junior to Griffin, then-Major Rich eventually was tapped to help form an initial cadre of hand-selected pilots that brought the F-22 Raptor fighter into operational service during the mid-2000s. Typical of smart, systems-oriented pilots who had grown up playing video games, Rich also had distinguished himself as a master air-combat tactician. He and his colleagues had developed novel ways to capitalize on the Raptor's cutting-edge "low-observable," or stealth, characteristics; nonafterburning "supercruise" speed-and-range combination, and sophisticated sensors and systems that were now integrated into Air Force network-centric-warfare doctrines.

As a result, Rattler Rich had been promoted, then selected for early advancement to brigadier general. That was the good part, he was fond of saying. The bad part was being pried from a Raptor cockpit and reassigned to a long-delayed "joint" tour, a prerequisite to pinning on his first star. He was currently overseeing a special-projects section for the National Security Agency director.

"Are you alone?" Rich asked. *What the hell?* Griffin thought, irritated at his buddy's cloak-and-dagger routine.

"Yeah, I am. The damned office door's shut. You want me to crawl under my desk, too?" Griffin half-smiled, recalling that Rattler had never given a hoot about such trivial conventions as military rank. As a one-star general, Speed outranked the bird-colonel, but you'd never know it by their frequent testy interchanges, a friendly banter honed years earlier within the close-knit fighter community of carefree bachelors.

Rich ignored the jab. "Look, Speed. Some really, *really* bad shit's happening in your backyard. Right there in the White House, and it stinks like a Baghdad back-alley. You're working directly for the skunk himself, so I've gotta get your take on this. It doesn't make a damned bit of sense, and I don't know what to do with it," he rambled in hushed tones. "But this *is* super-close-hold. Copy?"

Griffin glanced around, ensuring nobody could hear him. In the current administration's White House, military officers took precautions. "I copy. What in *the* hell are you blabbering about, Rattler?" he stage-whispered.

A long breath was drawn on the other end of the secure-voice link. "This is gonna sound nutso-crazy, but hear me out."

"Just spill it, numb-nuts!" Griffin said, louder than he'd intended. Rattler never pussyfooted around like this, and Speed was under an impossible deadline to finish an assessment for his civilian National Security Council boss, Paul Vandergrift.

"Keep your frappin' jockstrap on, Speed! This ain't easy. Listen, we have ironclad proof that your head honcho, the National Security Advisor and NSC Director, is communicating on a secure back channel with some Chinese People's Liberation Army spook. It sounds like they're making a deal, some kind of give-and-take exchange. But it gets worse. This sure as hell looks like *we* lose people, ships, and airplanes in the process. It stinks, man!" Rich spat.

"Whoa, *whoa*! You're saying Mr. Vandergrift's cooked up something with a Chinese *spy*? Sounds like grade-A spook bullshit, Rattler!" Still, Griffin mentally flashed to Vandergrift's recent missteps, several off-the-books, massive screwups that had cost the lives of two people at the hands of an Iranian madman and his Hezbollah proxies.

"Hell, I know it sounds ridiculous!" Rich barked, then raced on. "But I'm not shittin' you! We've voiceprinted both of these guys and cross-checked 'em. There's no doubt about who's talkin', and what these dudes are cooking up is scarier than hell. In fact, Vandergrift is . . ."

"Just a friggin' minute," Griffin interrupted. "Where'd you dig up this crap? NSA doesn't monitor domestic communications. Congress put the kibosh on that civilian phone-tapping program . . . what? Five or six years ago? Let alone comm coming out of *this* place!"

A long pause hung in dead air. "Speed, you're not cleared for the 'how,' and I can go to jail for even making this call. Suffice to say that I'm into supersecret, 'deep-black' doo-doo over here, okay? But I can tell you this: Under the Patriot Act, a special unit was set up here at NSA during the Bush-W years to monitor comm between high-ranking administration and congressional folks *and* foreign nationals. I don't know what set it off. The first National Director of Intelligence, back in '05, put the order in place, and it still stands. It's 'blacker' than the proverbial black cat in a well at midnight, but I guaran-*damn*-tee you it *did* happen and it *is* still active, no matter what the eunuchs in Congress dictated. I've seen the *Special Compartmented Memo* that dictates exactly who and what were to be monitored," Rich affirmed.

"Someone waaaaay up there in the Bush Administration must have suspected they had another 'Deep Throat' buried in the White House staff—a guy talking to foreign dudes, evidently. One of my jobs here is to oversee those communications with foreign entities. Usually, we just see bureaucratic, yawner-class nonsense, but not this. That's more than you're cleared for . . . *sir*, and you *never* heard any of this from me! Over!"

Griffin shook his head in disbelief. He hated Washington's dirty little secret games. *Geez! Nobody trusts anybody in this town!* he thought. "Copy all. You have a crappy job, Rattler. Do you sleep at night?" It was trite, but Griffin was trying to wrap his mind around the ramifications of what he'd just heard. Not waiting for an answer, he plunged ahead.

"Okay, let's say Mr. Vandergrift's talking to a Chinese guy. There *are* some really nasty political gyrations going on at the moment. You know that. And the PLA's just assumed a threatening readiness posture, thanks to a couple of our aircraft carriers showing up in their neighborhood. Plus that outa-left-field North Korean nuke blast— and God knows what else might be stirring 'em up. The President's preparing to send an envoy to Beijing, trying to defuse a very tense situation. At a time like this, I'd *expect* to see the NSC director working back-channel contacts, wouldn't you? The prez is bound to exercise every damned avenue he has to avoid a nuclear face-off. Frankly, I don't see a burning reason for concern, Rattler. You've been playing spook too long, bud. You're gettin' blued into the 'everybody's-a-spy' mentality."

"Damn it, Speed! Give a fighter pilot brother credit for a gram or two of smarts!" Rich exploded. "You know me, dude! Ol' Rattler's ten times more skeptical of spook-shit than you are! Listen, something very bad-serious is going down, and you've gotta help me out!

Shut your pretty face and . . . uh . . . Sorry . . . General, *sir*. But we're into some really nasty stuff here!"

Rich was deadly serious, Speed concluded. Rattler was one of the straightest arrows Griffin had ever encountered, a levelheaded, always-cool, combat pilot and commander, a guy Speed had never seen go off half-cocked.

"All right, all right! I'll concede you *might* have a half-brain, even though I'm skeptical about *this* wild gumshoe tale. So, shoot. What convinced you that Van's back-dooring us?" It was as close as a fighter pilot could come to apologizing.

Rich outlined the gist of secure-communication messages that had whipped back and forth between Vandergrift and a Chinese state-security type identified as Feng Bao Nu. Given that Nu was in possession of a U.S.-developed, scrambled-communications device compatible with one Vandergrift used, Griffin reluctantly conceded that Van probably was, indeed, arranging a deal with Nu. How that comm system had gotten into Nu's hands was a huge question, because the devices were tightly controlled. Only specific, high-level U.S. officials could get them. And it sure didn't sound like Van's proposal to Nu was anything the president or SecDef would have agreed to. Griffin fired a few probing questions, looking for holes in Rich's case. There weren't any.

"Let's see if I've got this straight," Griffin finally summarized, closing both eyes and squeezing the bridge of his nose between a thumb and forefinger. "Vandergrift's agreed we'll soak up a Chinese attack of some kind. In return, Nu will allow the U.S. to take out a laser-weapon site near Beijing, right?" *The lasers that zapped me on that last XOV mission! Christ . . . !*

During a highly classified Blackstar two-stage-to-orbit spaceplane mission, Griffin had maneuvered his XOV, or "experimental orbital vehicle," close to a constellation of Chinese nanosatellites, obtaining critical intelligence about the tiny reconnaissance spacecraft. While Speed was shooting high-resolution photos, his single-pilot spaceplane had been hit by high-energy laser beams fired from a terrestrial site in China. The lasers had weakened the XOV's already damaged canopy, which barely survived a gut-wrenching reentry. The memory of that life-threatening mission was only too fresh.

"That's it, dude," Rich confirmed. "And it's playing out as we speak. That's why I called you . . . *sir*. After that Korean nuke went off, the prez put B-1s and B-2s on alert in Guam, Okinawa, and Diego Garcia. Supposedly, he was sweating a possible attack by North Korea or China. Most people think the bombers would head for North Korea, if the prez says 'go.' Not exactly, bud. I did some digging and found that a few B-2s are primed and cocked to switch into hard-stealth mode and blast into China. And one of the targets is . . . ?" Rich asked cryptically.

"Ground-based laser sites," Griffin finished, nodding. "And if the B-2s turn those GBLs to scrap iron, the Chinese will have a reason to attack our carriers—or some other juicy target."

"You got it. Now, here's the scary shit: Nu's telling Van that a one-for-one trade is all that'll happen. That's it. One exchange and everybody stands down. Then peace reigns. Nu claims he's representing a 'moderate faction' inside the Chinese government, and his guys will step forward and demand an immediate halt to the craziness, before both sides go nuclear. Vandergrift does the same here, *and* gets brownie points from the president and all of us Washington-worshipping citizens for being the hard-ass tough guy who made the nasty ol' Chinese back down."

Griffin still couldn't get all the pieces to fit. "I hear you, and it resonates. Van's been badgering the prez to hit the Chinese before they slam us. In fact, he's specifically advocating taking out the ground-based laser site that the CIA thinks zapped me." Griffin memory-flashed to a cryptic NSC director comment made after Speed returned to Washington, following that dicey spaceplane mission. Something like, "Don't worry, General. China will pay for that attack."

But why would Van agree to a limited exchange that could easily escalate and spin out of control, perhaps triggering nuclear war? And how in the hell could Vandergrift be sure a B-2 would be launched against a specific target in China? Those bombers were under absolute positive control, particularly when carrying "hot" weapons. And if somebody on either side changed his mind or miscalculated how much power and influence he actually wielded, the whole screwy scheme would go south! *Way too shaky—absolutely crazy!*

"Why would the NSC director, of all people, be doing something so damned risky?" he asked rhetorically, thinking aloud.

"I was hoping you could tell *me*, Speed," Rich said evenly. "Is the President actually pulling the strings here, trying to build a tough-guy image for political benefit? If so, and he's dealing with a Chinese spy through Vandergrift . . . That's grounds for impeachment, at the least! That is, *if* the United States survives a li'l ol' nuclear exchange."

Rich's bombshell theory stunned Griffin. But it still didn't compute. *No way! Boyer's the dove of all doves! He'd never go for a cockamamie hawk scheme that could go nuclear!*

"I don't think so, bud. Boyer's lookin' for a way out," Griffin said slowly. "No; the president is clean, I think. Van's probably doing this on his own—another of his screwball backdoor gambits. I don't know why he's doing it, but that's my gut-feel. I need to think this . . ."

"Speed, forget the thinkin'-on-it crap!" Rich exploded again. Griffin heard the colonel slam a fist on something solid. "This is no time for navel-gazing, *General*! We have B-2s and B-1s on strip-alert in the Pacific, armed with live ordnance, at this very minute! In a heartbeat, some of those big boys could be hauling ass into China to hit that GBL site and who knows what else! Hell, they could be in-country before you or I know they've even been launched! Speed, we gotta do something to stop this nightmare!" It wasn't clear whether he was talking about stopping bomb-laden B-2 Spirit and B-1 Lancer bombers, or tackling a power-hungry, double-dealing NSC director.

Griffin's mind was racing, exploring options. There was no way he could waltz into Vandergrift's office and confront him with this. Who else would listen, then act in time? The president's chief of staff? He dismissed the idea before it fully formed. Confiding in Gil Vega would be personal suicide, while triggering alarms throughout the White House and the Pentagon. And Speed Griffin would be transferred to Timbuktu or Thule, Greenland, before sunset.

Who the hell would take a puny one-star's word over that of the president's National Security Advisor? Speed Griffin's sterling career would be instant charcoal. And confronting Vandergrift would drive the NSC director underground, neatly covering his political tracks in the process. Better to keep the SOB in the dark, so NSA's comm-interceptors could monitor what Van was up to.

"Look, Rattler, I'll take you at your word, and I'll do my damndest to keep those bombers on the ground. Of course, if you're wrong, we're both dead meat. If you're right . . ." The image of a mushroom cloud rising over Beijing and another over Washington played as a mental horror-movie preview. "Christ help us!"

The two officers reconfirmed their respective action items and signed off. Griffin sat quietly for several moments, then selected a speed-dial option on his handheld secure-voice communicator. There was only one man he could trust with dynamite like this. If *he* couldn't stop this madness, nobody could.

STRATCOM HEADQUARTERS/COMMANDER'S WORKOUT ROOM

Sitting on a padded workout bench, General Howard Aster snapped his handheld communicator closed and tossed it on the recycled-rubber flooring, next to a half-empty water bottle. He yanked a short terry cloth towel off the bench, mopped his face, then twisted the towel, staring at nothing. The officer wore a gray-and-blue, sweat-stained Air Force T-shirt, dark-blue, knee-length cotton shorts and size-twelve Nike running shoes. A barbell loaded with 180 pounds of paint-dinged, circular weights rested on twin supports, suspended a few feet above one end of the bench.

Across a compact workout room crowded with racked weights, horizontal and angled benches, a universal, all-in-one exercise machine, and a treadmill sporting a well-worn belt, Burner Burns stood with his back to Aster. The lieutenant colonel exhaled in loud, periodic bursts as he curled twin 70-pound dumbbells, one in each hand. Rivulets of water ran off bulging and rippled shoulder muscles exposed by a skintight tank top. Burns faced a full-length mirror, which allowed him to monitor both his own lifting form and his boss, the STRATCOM four-star commander. They were the only occupants of the tiny gym.

The general's early morning workout routine had been interrupted by a call from the east coast, a call that obviously had disturbed Aster immensely. He'd ended it abruptly, then immediately dialed the Secretary of Defense at the latter's office. But an air-conditioning vent in the ceiling had kicked into action as Aster spoke to Hurlburt, creating enough noise to preclude Burns catching much of the conversation.

Something sure screwed up the boss's morning, Burns observed. Aster sat quietly, flicking the towel aimlessly, his thoughts miles away. Burns had heard enough to know the telecon's subject involved the National Security Agency and National Security Advisor Paul Vandergrift. And that smug political hyena Vandergrift usually meant trouble. A call from Washington this early in the day always spelled more work for Aster, STRATCOM, and, inevitably, Burner Burns. *Too bad. The boss already has too much dog-doo on his doormat.*

Burns watched Aster via the mirror, raising, then lowering, the barbells, first the left, then the right, stressing his biceps until they quivered with pain. He waited. Aster would say something, *if* the general wanted to talk.

Snapping the small towel sharply, Aster mentally replayed the two phone conversations, again wondering how the hell the U.S. had gotten into such a mess. He was emotionally tired and discouraged, a rare combination for him. Normally, a rigorous morning workout in the private, makeshift gym down the hall from his office reinvigorated his body, cleared the mind, and brought new insights to knotty problems. Not today.

The staggering info about Paul Vandergrift and Feng Bao Nu that he'd fielded from Speed Griffin, then relayed to Secretary of Defense Hurlburt, was potentially deadly from two aspects. If incorrect, the serious charges leveled at Vandergrift *would* scuttle Aster's career, as well as those of two other up-and-coming senior officers. If correct, and both he and Secretary Hurlburt were unable to stop the madness unleashed by a friggin' Washington traitor—Aster couldn't help but consider Vandergrift as such—it could unleash a nuclear nightmare unlike anything Hollywood or imaginative Pentagon wargamers could ever dream up.

Aster knew exactly how a military strike on China would play out. His wargamers had examined that particular scenario from multiple angles shortly after Speed's spaceplane had taken two laser hits, and the outcomes were always the same: U.S. retaliation against that Chinese ground-based laser site would spark a rapid tit-for-tat military escalation, which couldn't be arrested. A U.S.-China nuclear exchange would be highly probable, his gamers had repeatedly determined. That outcome had been reported to President Boyer, the SecDef and Admiral Lee, who was running a side wargame focused on China.

Hurlburt had pretty well summed up the situation a few seconds ago, as he'd signed off. "If Van believes this Nu sumbitch can single-handedly stop the PLA from launching a massive counter-attack, he's out of his warped mind! This *will* go nuclear, Howard!"

OVER THE ATLANTIC OCEAN

Hassan Rafjani roiled a snifter of fine cognac, took a sip, then closed his eyes and enjoyed the rich warmth bathing his throat. Yes, drinking alcohol was forbidden by Islam, but powerful Iranian figures were exempted from such little-people constraints, at least while outside Persia's borders. Especially while winging across an ocean at 40,000 feet-plus in a whisper-quiet, luxurious Dassault Falcon 7X business jet. After an intense flurry of high-level meetings in Caracas, Venezuela, he was mentally exhausted. The amber liquid soothed both his weary mind and tense muscles.

He opened his eyes when Iran's president settled into a facing seat. The swivel-chair's soft, cream-colored leather seemed to swallow the small, bearded man. Mahmoud Ahmadinejad crossed bird-like legs, smiled, and raised his own glass of cognac. Dark eyes narrowed to slits.

"A very successful encounter, yes?" the president said quietly, assuming his trademark reptilian air. Rafjani returned the raised-glass salute.

"Very much so," Rafjani replied, taking a sip. Ahmadinejad stared out a porthole-like window a few moments, before speaking softly.

"Can we be assured that these new allies will, indeed, deliver the oil, gas, and food they've promised? And carry out the space mission they claim is prepared to launch soon?"

Rafjani hesitated for effect, gently swirling the cognac and also staring through the jet's tinted oval windows. Far below, the water's undulating surface caught and reflected flecks of the Sun's late-day light. So distant, yet calming.

"Chavez and Zubkov are both emotionally invested in this arrangement," he said finally, referring to Venezuela's dictator and Russia's caretaker president. The latter had been an obscure financial regulator, before being elevated to the high-profile position by the previous president, Vladimir Putin. However, most knew Putin and his cabal continued to dictate Russian policies as the actual behind-the-scenes power base.

"It is in their *personal*—not only their nations'—interest to uphold their end of our bargain and deliver these essential goods to our friends in North Korea," Rafjani added. "Be assured, my brother, that Kim's atomic weapon blast has greatly upset both Russia and China. The North has been brutalized by its so-called 'sponsors' for that 'brash' act. And yet, both Russia and China now are acutely aware of North Korea's desperate economic state, and are much more likely to respond to the North's pleas for assistance than they were before the high-altitude detonation. Fear is a powerful weapon, and Kim's act has frightened our new allies tremendously. I believe we will see Venezuelan and Russian oil and food flowing to North Korea very soon, as a way to pacify Kim and assure the West that the North is 'under control.' "

Ahmadinejad nodded, reflecting silently that he was looking at one of Iran's most powerful, yet virtually unknown figures. America's storied CIA knew virtually nothing about Hassan Rafjani, nicknaming him "Dagger." The CIA's file on the man was thin, at best. Of course, they were very aware that Dagger had slowly skinned an undercover CIA agent with a razor-sharp, rippled blade, before killing the terrorized infidel.

A fitting moniker, Iran's president mused. *I must keep this dangerous being close. He would be a formidable enemy.* Aloud, he said, "I marvel at your persuasive powers, brother. Convincing the North's fanatics to launch and detonate their one and only remaining nuclear weapon was masterful. That singular act has advanced our timetable substantially. The Islamic Republic is most indebted." Again, the president raised his glass in salute.

Rafjani smiled, accepting the compliment for what it was. He and Ahmadinejad had been wary colleagues since the 1979 revolt that had ousted the Shah of Iran and established a ruling theocracy now controlling most of the nation's formerly secular society. The two had been mutually useful, and gave the outward impression that they were close friends and political allies. They both nurtured that perception, but it was far from the truth. Neither trusted anybody, least of all each other. Both were masters of manipulation, though, and had agreed to cooperate. Their odd symbiosis had been quite successful—and mutually beneficial. Working with the ruling mullahs, Rafjani had ensured Ahmadinejad's rise to power on the political side of Iran's ruling class. In turn, the Iranian president promoted Dagger's influence on the theocratic side of the government by openly turning to Rafjani for counsel. As a result, Rafjani wielded enormous national power, much of it exercised through the crafty, engineer-trained Ahmadinejad.

The president nodded, still smiling. "Then Kim will be satiated and indebted to us for the forseeable future?"

"Certainly. With desperately needed food and oil flowing from Venezuela and Russia, Kim will be less susceptible to economic pressure from China. He can assume a much harder, more independent stance—and will respond to your 'suggestions.' And, as a result of Kim's brave atomic spectacular, the Great Satan must turn its attention to North Korea, ignoring our beloved Persia—at least for a while," Rafjani summarized. "As you know, many of Satan's spy satellites are dead, damaged, or dying from the blast's atomic radiation—and, prior to that, from our Colombian friends' maser, of course. America can no longer monitor our activities closely, which offers us a brief window of time to complete Allah's Sword without the Satan's interference."

Rafjani had some lingering doubts about the current state of America's radiation-hardened intelligence-satellite network. But his scientists had confirmed with great confidence that many commercial communications spacecraft were damaged or completely disabled. And since the United States' military forces relied on commercial comsats for roughly eighty percent of their long-haul communications needs, the Great Satan's formidable combat capabilities had been crippled severely.

"We must complete the campaign soon," Ahmadinejad declared. Rafjani nodded, taking another sip of cognac. The president again turned to the oval window and sighed deeply. "You have prepared well, my brother. But the time to strike is upon us. I know you have been occupied, implementing the first thrusts of 'Sword,' so I am compelled to share what *I* see happening to Persia of late. Our people are becoming increasingly restive, demanding lower prices for gasoline, food, housing, and other essentials. The national treasury is virtually empty, and we can no longer afford to subsidize gasoline and food prices to keep them artificially low. Soon, I will be forced to ration gasoline even further, because the Europeans and Arabs have suffered from the global economic crisis and will not extend additional credit. That prevents us from importing desperately needed fuel. Our people are growing angry, and will no longer accept that outsiders are completely to blame for their misery. Now, they blame us, their leaders."

He turned from the window and held Rafjani's stare a moment. "Iran is on the verge of revolt. Unless the people are given a dramatic reason to unite and resist invaders who would harm them, our Islamic Republic will fall, a victim of its own citizens' wrath."

"But our brothers *know* that the Great Satan and its slaves are responsible for their pain! America and the Europeans are to blame for choking off imports with their evil sanctions!" Rafjani said heatedly. "Our people see the arrogance of these infidels! If not, then they *will* understand, when the enemy feels the bite of Allah's Sword!"

"Ah, yes; most will. And very soon." Ahmadinejad laughed softly, amused by his partner's dark fury. Dagger's anger was so easily ignited. "Indeed, your twin strikes in California were excellent previews of Sword's fury! Brilliant diversions, my brave Dagger! But the time for Phase Two is now. It cannot wait. Are you prepared to unleash Allah's final blow immediately?"

Rafjani struggled to control a white-hot hatred for the West, a deep-seated anger that, of late, threatened to consume him. He pulled a long draught from his snifter, draining the glass. The fiery liquid worked its magic, soothing and restoring a semblance of control to his tone. "We are ready," he replied after an uncomfortable silence. "The airliner and containership incidents were mere pinpricks. As a result, the infidels are flailing about, cringing in fear of yet another strike on their perverted soil. As if a thousand deaths of American swine matter to God!" He threw his head back and laughed loudly. Leaning forward, Rafjani lightly touched the Iranian president's knee with a long, bony finger. "Allah's Sword is ready and waiting for its final thrust. My trusted Quds in the rocket forces stand ready to restore Persia's glory. We only await your 'go' order, my brother."

Ahmadinejad studied the man before him, no longer smiling. Black eyes bored into Rafjani, searching Dagger's soul for . . . what? Compassion? Love of life? A touch of humanity? He saw nothing but a seething hatred smoldering in the dead space that once had held the gaunt man's soul. *How can one truly serve Allah, when there is no light of life inside?* the president wondered absently. Rafjani's eyes burned, never wavering, staring back. Ahmadinejad tilted his head, lips spreading into that familiar slit-eyed smile.

Another long silence hung between the two men as they eyed each other, neither blinking. Ahmadinejad finally raised his glass slowly and took a sip, before speaking. "I must ask again. Do you trust the Russians and our new friend, Chavez, to fulfill their commitment in space? Can they really destroy the new American outpost in orbit?"

"Yes. Without question," Rafjani said confidently. He settled back into the leather seat, palms pressed flat against his bony thighs. "I reviewed the entire plan this morning, while you and Mr. Chavez were speaking to the CNN cameras. The Russians assured me the satellite is ready, poised to launch as we speak. Chavez's engineers also assured that, once the vehicle is in orbit, they are ready to carry out its mission. Within days, it will strike yet another blow for Islam!" He smiled widely. "And the infidel fools believe it is a research spacecraft!"

"Then go. Launch the final solution," Ahmadinejad said sharply. Rafjani's smile widened, a tingle of excitement racing through his body.

"History will record this as Persia's greatest hour, my brother. Your bold leadership will be the subject of legends repeated for generations. Never again will Islam be shamed by the Crusaders and their Zionist puppets!" Rafjani's voice was husky with emotion, his eyes shining. Ahmadinejad nodded in mock humility, then stood, raised the snifter once again in salute and returned to his private quarters.

Rafjani's body quivered with anticipation. He mentally reviewed the next steps of Allah's Sword, noting that, in a matter of days, the world would be a much different place. The fires of Hell would sear the desert, cleansing it of the vermin who had defiled God's Earth far too long.

OFFUTT AFB, NEBRASKA, FLIGHT LINE

General Howard Aster opened the staff car's door and unfolded his lanky frame, not waiting for his driver, Burner Burns, to open it for him. Wearing a tailored green flight suit with four black stars embroidered on each shoulder, a silver-edged blue flight cap tilted forward until its edge almost touched the bridge of his nose, the STRATCOM commander was the quintessential image of a confident fighter pilot.

Aster held a hand on the flight cap, crushing it against the top of his head. He would always be uncomfortable wearing the damned cap on a flight line packed with aircraft. He'd spent years automatically tucking the envelope-like headgear into a flight suit's zippered ankle pocket as soon as he headed for his F-16 fighter, carrying a helmet bag stuffed with a "brain bucket," kneepad, and other gear. Flight caps were simply a likely source of FOD, or "foreign object damage," a name given to anything that could be sucked into a screaming turbine engine, catastrophically destroying its hundreds of thin blades in seconds. But generals who met arriving VIPs were expected to be in full uniform, even on a flight line, and that meant being "covered," wearing a uniform hat.

A few quick steps placed him near the door of a sleek, white-painted C-21, the U.S. Air Force's version of a Learjet 35A. The business jet's drop-down door with integral stairs descended, even before the twin Honeywell/Garrett TFE-781-2-2B turbofans whined to a stop.

A burst of golden hair topped by another blue, silver-lined flight cap ducked through the door. When she looked up, Major General Dawn Erikson's smooth, yet angular features broke into a broad smile. She stepped off the lower step, popped a quick salute, and extended an open hand.

"General Howard Aster! It's so good to see you," she beamed. He was struck by her perfect white teeth, framed by a model's ideally shaped lips, and firm handshake. *Holy shit! She's still gorgeous!* he flashed.

"Howdy, Dawn. It's damned good to see *you*, too! Hey! Double stars! Last time I saw you, those were silver leaves!" Aster replied, holding her outstretched hand an extra-long beat. Although she was tall for a woman, Aster took note of how petite she seemed. Her soft hand disappeared in his. She wore an olive-drab, standard-issue flight suit just snug enough to reveal a trim, shapely figure. Polished black jump boots were incongruously unfeminine, but were part of the uniform.

"That's right, sir. They were." Erikson's pale ice-blue eyes sparkled as they searched his, apparently finding what she wanted.

"You look terrific, Viking," Aster said huskily, adding, "And I'm damned glad you're still alive and in one piece!"

She started to answer, but noted Aster glance over her head. Army Major Hans Richter, her aide, was ducking through the C-21's door. Aster's executive officer, Burner Burns, was there, offering a hand to the stocky major, as the latter wrestled with a set of crutches.

"Thanks, sir," Richter said, accepting the man's assistance. "These blasted sticks didn't come with instructions." He tucked the crutches' padded ends under each arm and gingerly placed their rubber-capped tips on succeeding steps of the C-21's door, lowering his weight to each in turn. Burns grasped the man's left bicep, steadying him until Richter reached the concrete tarmac.

"Howard Aster," the STRATCOM chief said warmly, offering his right hand. "It's an honor to meet you, Major. How's the leg?" Richter leaned on the crutches, and took the general's hand.

"Major Hans Richter, sir. It's an honor to meet you, as well. General Erikson's told me a lot about you, sir. And thanks; the leg's fine. No serious damage."

Aster grinned, glanced at Erikson, and replied, "Well, I hope you didn't buy all of Viking's tales! Especially about our knock-down arguments during Gulf War Two."

Richter looked to his boss, not sure how to respond. Erikson dismissed her aide's uncertainty with a wave and turned to Aster, hands on both hips. "Sir, that was a long time ago. Besides, you were right . . . *most* of the time." She smiled and tipped her head toward Richter. "General, you've just met one of the bravest men I've ever known. Hans saved a lot of lives in L.A., including mine." Richter smiled broadly, but dropped his eyes, embarrassed.

"Damn straight! The nation's indebted to you, Hans! If not for your taking control of the situation, that could have been a nasty bloodbath," Aster said, slapping the Army officer on a shoulder. Shifting his attention back to Erikson, Aster's tanned, lined features broke into a lopsided grin. "And I hear the *L.A. Times* dubbed *you* Annie Oakley-in-blue, Viking." She smiled and again flicked a small hand dismissively.

Aster herded the guests toward his staff car, opening the front door for Richter, while Burns held the right rear one for Erikson. A crew chief loaded a couple of roll-on suitcases into the trunk, and handed Richter two slim briefcases. Aster slipped into the back seat beside Erikson, as Burns climbed behind the steering wheel. Chronic budget crunches and personnel shortages had prompted Aster to forgo a full-time enlisted driver, tasking Burns with yet another extra duty.

As Burner drove slowly across the broad expanse of Offutt Air Force Base's concrete flight ramp, Erikson looked up at the silver-haired officer next to her. "Thanks for sending that C-21 for us, sir," she said quietly. "Once the feds closed all airports in Southern California, I didn't think we'd ever get out of L.A.!"

"My pleasure, Viking. Of course, I had my own selfish reasons, you know," he smiled. She returned it, vaguely hoping a double meaning was camouflaged in Aster's statement. "Any trouble getting up to Edwards?"

"Not really. The protocol folks at Space and Missile Systems

Center were terrific. They jumped in and rescued us from the news media, helped me hijack Hans from a local hospital, then drove us up to Edwards. That may be our top-notch Air Force flight testing base, but I really didn't relish the thought of being stuck in the desert for days. Edwards is getting hot this time of year, and the wind *always* blows, of course. Your C-21 saved this girl from terminal bad hair, sir! I can't thank you enough."

Stealing a glance in the rearview mirror, Burns thought her pale blues held Aster's a tad too long. The boss was eating it up, though. Two accomplished senior flag officers grinning like schoolkids on a first date! *What the hell . . . ?*

Burns glanced at Richter, wondering if the Army officer also had picked up the sparks and chemistry between the two generals, but the bald major was eyeing Offutt's flight line, scanning left and right. Richter might be General Erikson's designated aide de camp, but the guy acted more like a Secret Service bodyguard. Eyes always moving, searching, his entire being coiled and ready, it seemed. Nothing escaped the big Army Ranger. That heads-up situational awareness had saved Erikson's and Richter's lives, Burns reflected admiringly.

Under Aster's warm gaze, Erikson suddenly felt very exposed. A flush of warmth and a slight tingle rippled through her. She recognized, but tried to suppress, an old, familiar excitement mixed with longing. She looked away, concentrated on steadying her breathing and fought to ignore a delicious, yet painful, memory that threatened to flood her consciousness.

"Well, sir. Exactly why am I here?" she asked, trying to project only professionalism. "I was told you needed our help ASAP."

Aster sighed, pulled off his flight cap and slapped it across a flight-suited thigh. "I do, Dawn. I'm sure you've been tracking the status of our damaged, limping, screwed-up space infrastructure, so . . . *What?*" he asked, frowning. Erikson was shaking her head vigorously.

"No, I haven't. Of course, you'd think the National Security Space Institute—America's one-stop space-education schoolhouse— would be read into our current space posture, but we're not. I'm having a devil of a time getting timely info for NSSI, sir," she said. "We *should* be receiving classified reports as soon as they're available, but NSSI isn't in the loop. We're treated like space mushrooms, left in the dark!"

Aster was confused. "Hey, you report directly to Buzz Sawyer at Space Command, so what the . . ."

"At one time, we did, sir," she interrupted. "But not now. Despite General Sawyer's and my vigorous resistance, the Air Staff transferred NSSI to Air Education and Training Command. The Air Staff sees NSSI as just another interservice school. Consequently, my staff and I no longer get daily intel and space-infrastructure status reports from Space Command. And that's a *massive* hindrance to our mission! We can do a hell of a lot more than train people on the fine points of orbitology and rockets!" She smacked a small fist against her knee, and brushed a lock of hair from a now-flushed forehead. The lady was boiling, Aster thought, suppressing a grin.

"We're *supposed* to be the Air Force's primary, forward-looking milspace think tank, and we can't even get a damned on-orbit status report from the 'Jay-Spock' at Fourteenth Air Force!" she ranted.

Based at Vandenberg Air Force Base, California, 14th Air Force was the blue-suit space command's operational arm. Its five wings of officers and enlisted teams, backed by thousands of contractor experts, provided launch services, controlled dozens of national security satellites, and conducted defensive and offensive "counterspace" operations, which ranged from tracking roughly 19,000 pieces of debris and space junk that cluttered low-Earth orbit, or LEO, to ferreting out the source of communications-satellite jamming. The heart of those day-to-day activities was the Joint Space Operations Center, known in military jargon as the "Jay-Spock."

"Okay, okay! Copy all," Aster said, raising a palm. "That makes absolutely no sense for the Air Force's one and only national security space think tank. But let's work that issue later, Viking. Right now, we have some serious space policy and doctrinal wolves biting our butts *here*, and *that's* why I need your inputs."

Erikson's jaw tightened, and that glare could have frosted the car's windows, Aster noted. But she said nothing. Aster shot her a lopsided grin . . . and she melted inside. That same sexy grin had disarmed her too many times throughout those long, tense days and nights in Qatar's Air Operations Center, during the last Gulf War. *Damn you, Howard!* she silently screamed. *Why do you* still *have to be so blasted good-looking?*

Aster quickly outlined damage suffered by the nation's military, civil, and commercial satellite infrastructure, painting a grim picture of reduced space-based capabilities. Weather forecasting was hindered by degraded imagery from satellites, thanks to damaged charge-coupled detectors and staring-infrared chips. Several highly classified signal, radar, and imaging intelligence-gathering spacecraft had been disabled by a drug cartel's maser in 2010, then others had taken hits from the recent North Korean nuke blast. Communication birds had suffered, but GPS satellites, for the most part, had ridden out the nuclear event in good shape. However, the GPS constellation was still limping from the earlier maser hits on a half-dozen Navstar spacecraft. Those were responsible for inaccurate position and timing information being beamed to Earth by a few GPS birds.

"Overall, we're hurting," Aster summarized glumly. "Buzz Sawyer's troops are scrambling to reconstitute some of the most critical capabilities, but the nation's financial crisis has dried up his budgets. Besides money, you know our other problems: no 'strip-alert' rockets sitting on the pad, waiting to fire new satellites back into orbit. And damned few spacecraft in the barn *to* put up, even if we had spacelift available!"

"What about our allies and . . . the others?" Erikson asked.

"Not good, either. The Brits lost several of their Skynet birds to Pyongyang's nuke, so they're hurting for comm support throughout the Middle East, Africa, and Afghanistan. Most of Europe's satellites are dead or dying, India's are in bad shape. 'Bout everybody who has something in space has been affected."

"Russia and China?"

Aster shook his head. "Still working on that. From voice and data chatter the spooks have picked up, it appears that Russia's situation pretty much mirrors ours. Aging birds were hurt most by the nuke, but newer vehicles appear to be in pretty good shape. China? Hard to tell. They're not talking, of course, but the NSA and Space Command's intel teams are fairly certain China's Yaogan remote-sensing satellites are out of commission, because they weren't radiation-hardened."

Erikson's brow furrowed. "Those are all Sun-synchronous radar-imaging satellites, right?"

"Right. But we know they're also equipped for visible-light imaging and reconnaissance. Otherwise, why put 'em in Sun-sync orbits?" Aster was again searching Erikson's eyes, looking for something.

The two-star nodded absently, thinking. "If they've lost all their Yaogans, do the Chinese have anything else up there to keep an eye on North Korea?"

Aster hesitated. "Well, they *had* a highly classified constellation of imaging nanosats in low-Earth orbit, and Buzz's intel troops think those birds probably survived. They were new platforms and had some—let's call them 'robust' characteristics." The general squirmed uncomfortably. Erikson was intimately familiar with that nervous Aster-response and instinctively pressed the four-star. "Sir, exactly what do you mean by *'had'* a constellation of nanosats?"

Aster shot her an irritated grimace. "They're still in orbit, but out of action, thanks to a mission I authorized. Without the gory details, Viking, we 'slimed' the damned things' optics."

Erikson stared at him for long seconds. Aster was familiar with that probing, icy gaze. The woman had often searched his being that way, during the time they'd served together. Those eyes were just as unsettling now. Dawn Erikson could peer deep into his very soul, it seemed. And he damn sure never liked feeling soul-naked. Hell, he was a nerves-of-steel, combat-honed fighter pilot, a leader of thousands, a man who routinely conversed with the President of the United States and Secretary of Defense! And yet, this slip of a woman could peel away his armor of emotional defenses and confidence, merely through her close proximity. She may be an Air Force flag officer, but Viking was an enchanting, lovely witch, as well. Her two stars—and his own four—were reduced to insignificance when she looked at him that way. At this moment, he was nothing but a raw-emotioned man, and she an enticing woman who could twist his heart at will.

"General, mind if I speak freely?" she asked softly, yanking Aster back to the now. He nodded, holding her gaze. "Sir, I think that 'sliming' mission was a big mistake. It'll bite us in the butt, big-time."

Aster flushed, his lips tightened in a thin line. "And exactly how did you come to that conclusion so damned quickly, *General*?"

She half-smiled and placed a soft hand on his, a move both natural and familiar. "My NSSI guys worked through a similar scenario as a cause-and-effect exercise recently. We were drafting a new space-control doctrine at the Air Staff's request, and that led to a heated discussion about the pros and cons of 'blinding' an adversary's recce satellites."

"And?" Aster asked peevishly.

Erikson squeezed his hand. "Sir, I'm sorry, but your folks made a big mistake 'sliming' those nanosats. I know I don't have the full picture, but I suspect we've literally blindfolded the Chinese. They don't know what the North Koreans are doing, and they don't know what we're doing. That loss of time-critical situational awareness is a classic formula for dangerous paranoia and miscalculation."

"Son-of-a . . . ," Aster grimaced. "That's what Admiral Lee's China 'Red' cell came up with, as well. After the fact, unfortunately. I'd already green-lighted the mission." He shook his head and swore under his breath, absently noting that Burns was pulling into the commander's parking spot in front of STRATCOM headquarters.

"Dawn, from a strategic, force-protection viewpoint, sliming those nanosat SOBs made good sense. For a number of reasons, it didn't feel right, I admit. But I didn't have a solid reason to say 'no.' You've reinforced my gut-level reluctance, and made a case for why I should have rejected that sliming mission." He donned the flight cap, and with practiced smoothness, crushed its top, forcing the aft edge to stand straight up in a rooster's tail. The front crease formed a "V" between his eyebrows.

"Maybe I screwed up. Maybe not. Either way, as far as China's concerned, I'm afraid we just took a giant step closer to the nuclear abyss."

8 CONFLUENCE

WHITE HOUSE SITUATION ROOM

President Pierce Boyer scanned three faces around the table—Secretary of Defense Hurlburt, the four-star Chairman of the Joint Chiefs, and Paul Vandergrift, head of the National Security Council. All were tired.

"Proceed," Boyer directed.

Vandergrift began, "Mr. President, you'll recall some prior intel regarding a Russian and Venezuelan collaboration on a Latin American space venture involving our friend Chavez."

Boyer nodded.

"Well, the collaboration was a lot further along than we realized. We have confirmation that Chavez's visit to Russia will be highlighted by the acceptance of a Russian-built satellite . . ."

"When?" Boyer interrupted.

"The transfer is imminent. 'VenezSat' will probably launch, within hours, from Baikonur, Russia's traditional launch facility."

Boyer leaned forward. "Did we expect this? At least, so soon?"

Hurlburt responded, "Sir, the CIA is working on a National Intelligence Estimate that updates the entire Venezuelan situation, including this space foray."

"Which has been in the works forever," Vandergrift interjected. "We asked for that NIE right after we came to Washington!"

When we *came to Washington! Arrogant bastard*, Hurlburt fumed silently.

Although the intelligence community's disparate agencies ostensibly had been united under a single directorship, individual units still jealously protected their turf. As a result, the NIE, once a capstone document indicative of significant trends and capabilities of America's adversaries and global interests, had become largely meaningless pablum. It had been reduced to a stratified document merely reflecting the lowest common denominator of consensus-by-committee, another vehicle for blatant pandering to powerful politicians and a fickle media.

Hurlburt tried again, ignoring Vandergrift's gratuitous jab. "Mr. President, our military attachés have developed very strong relationships within the Venezuelan high command. They assure us, with confidence, that Chavez hangs by a thread."

"Same rumors we've heard for years," Vandergrift scoffed.

"But what *you* don't know is the first major sacrifices of Chavez's failing socialist agenda are finally surfacing—and they're painful," Hurlburt continued, scowling at the NSC director. "Senior generals are worried that paychecks for their troops will soon become intermittent, at best, and maybe even cease entirely. Like all delusional communists and socialists, Chavez believes he can order his military to show Latin honor by setting an example: work for no pay. Remember their oath, 'Live for Chavez, die for socialism.' But rumblings among some generals and many colonels have definitely begun. A coup wouldn't surprise any of our intel folks."

"Okay; I get it. What else?" sighed Boyer.

"The string of economic failures throughout Venezuela is staggering. Ya gotta love the twisted logic that always appeals to the masses," the Secretary of Defense added. " 'Take from the rich and give to the poor' never fails to resonate among the have-nots. Nothing more than a Ponzi scheme, run by the government, no less!"

The president shook his head and slapped the table. "Hold it, hold it! That's very interesting, but help me understand the relationship to this imminent space launch."

"One, the launch may be a diversion, Mr. President," Hurlburt said. "But it may also be a Cuban redux—the 1962 missile crisis in a new cloak. Chavez needs the Russians for legitimacy, but he may also be jockeying for plain old geopolitical-thuggery reasons. Iran provides political fervor and money, but does nothing to prop up Chavez's power base."

"Then why this expensive space venture?" Boyer asked.

There were no answers. Not in that room, at least.

BAIKONUR COSMODROME/KAZAKHSTAN

Standing on a pillar of white-hot fire and thick, billowing smoke, a Russian Starsem Soyuz rocket slowly lifted off a scarred, fire-blackened launch pad and accelerated into a pale-blue sky. The rocket's deep-throated crackle rolled across the aging launch site, reverberating against the chests of Russian and Venezuelan officials crowded onto a spectator tower several miles from the pad. Necks craned, and most shaded their eyes with cupped hands to keep the flame's rapidly shrinking spot of brilliant white in view.

Venezuela's president-for-life, Hugo Chavez, pointed at the diminishing plume and chattered wildly, challenging an interpreter to keep up with his litany. Standing beside Chavez, Russia's prime minister, Dmitry Zubkov, smiled and nodded proudly, reveling in yet another demonstration of Mother Russia's spacefaring superiority. He also took quiet pride in the knowledge that it was *he* who had made this moment possible.

While serving as an obscure finance official buried in Moscow's bureaucracy, Zubkov had proposed creating a new oil cartel to challenge OPEC, the Organization of Petroleum Exporting Countries. New revenues of almost $700 billion had flowed to OPEC nations in the late-2000s, then skyrocketed into the trillions as oil topped $140 a barrel before retreating. Zubkov had caught the eye of Russia's then-president Vladimir Putin, who elevated the finance bureaucrat to prime minister and ordered him to form a new oil triumvirate.

Within a few years, Zubkov had combined the petroleum resources of Russia, Iran, and Venezuela under a three-nation umbrella known as the People's Oil Consortium. True, its output was dwarfed by that of the Middle East-dominated OPEC, but POC was funneling billions of new dollars into treasuries controlled by three powerful leaders. And, to a person, those leaders hated the United States and its lapdog allies in Europe and Asia.

Thanks to Zubkov's skilled manipulations, first Venezuela's Chavez, then Iran's Mahmoud Ahmadinejad bought into the new cartel's most potent reason for being: employing oil as a powerful political weapon. Of course, POC had delivered far more income to Russia than to Venezuela or Iran, but the latters' fanatical leaders cared less about revenues than they did about furthering their socialistic agendas, while also foisting economic and political pain on capitalistic societies. And oil-leveraged pain was definitely being felt in myriad capitals, from Washington and London, to Sydney and Tokyo.

Zubkov squinted against Kazakhstan's bright sunshine until the rocket flame disappeared. He clapped Chavez on the shoulder, interrupting the still-chattering Venezuelan. Glancing at an interpreter, Zubkov said, "Mr. President, your country will soon have one of the world's most sophisticated research satellites in orbit! Congratulations!"

Chavez's response, in Spanish, caused the interpreter to blanch, but she dutifully translated his words into Russian. "The president says, 'Research my ass, comrade! Your loyal scientists and engineers have crafted a knife I will thrust deep into the arrogant Americans' heart!' "

Zubkov winced, but nodded politely. "Tell President Chavez to please avoid such inflammatory remarks. There *are* television and newspaper reporters here! We also must be careful to not 'telegraph' the strategy too soon. To be effective, surprise is very important."

Chavez flipped a palm skyward and shrugged, dismissing Zubkov's admonishment. He turned to his entourage and extended both arms, smiling broadly. "Today, my people," he shouted in Spanish, "Venezuela has assumed its rightful place as a new space powerhouse! This 'research' satellite, built and launched by our Russian friends, will uncover many secrets locked in the ocean of space above. Those secrets will no longer be controlled by a few perverted capitalists, who have selfishly conspired to harvest the riches of space for their own purposes. Venezuela will *share* these secrets with all people, bringing riches to the poor across the globe, just as we have provided oil to those abandoned by heartless capitalists!"

Although numbering less than a dozen, his Venezuelan hangers-on shouted long and loud, pumping fists into the air. Zubkov listened to the interpreter, exchanged a knowing look with a nearby Russian rocket scientist, who subtly shook his head. *The buffoon has no idea that he also insulted Russia, Earth's first spacefaring nation,* Zubkov thought wryly. *Maybe 'space secrets' should be kept secret from crazies such as Chavez.*

No matter, though. Chavez's $253-million check was in a Russian bank, so Zubkov's role in this caper was ended. As soon as that VenezSat reached orbit and was declared operational, Russian engineers would transfer full technical control—and *all* responsibility—to a Venezuelan team at a Caracas-based ground station. Whatever happened thereafter was Chavez's concern. Zubkov and Mother Russia would have no part of *that* nightmare.

AIR FORCE SPACE COMMAND HEADQUARTERS/ COMMAND CENTER

General Erik "Buzz" Sawyer was aggravated and getting more so as the daily commander's update briefing progressed. On a wall-size plasma display, the image of a colonel at 14th Air Force's Joint Space Operations Center in California delivered a dispassionate rundown of facts about a new Venezuelan-owned satellite launched by Russia from Kazakhstan. U.S. and Australian radars and powerful ground-based telescopes had tracked the launch vehicle and imaged a small spacecraft being deployed in low-Earth orbit. The satellite had been assigned a designator and added to the "space catalog" of roughly 19,000 man-made objects in orbit.

"Excuse me, Jay," Sawyer interrupted. Halfway across the country, the colonel stopped in mid-sentence and turned to a tiny videoconference camera. Sawyer, at Air Force Space Command's Colorado Springs headquarters, continued, "All those numbers and stats are great for engineers and guys who fly satellites, but they don't help *me* one damned bit. I need to know what that VenezSat is doing up there. What's its mission? Why was it launched into that particular orbit? Do we know *anything* about its payload and capabilities?"

The colonel turned and called up another PowerPoint slide before answering. "Sir, it's a research platform. Based on standard Russian pre-launch notifications to us and other nations, as well as information put out by Venezuela's space agency for news organizations' consumption, the satellite's designed to gather data on Earth's atmospheric constituents, and to check out a suite of new sensors optimized for 'locating natural resources.' "

Sawyer scowled and slapped the conference table loudly. "I don't give a hoot 'n hell what the Russians and Chavez's propagandists are putting out, Jay! I want to know whether that little devil has optical or radar sensors. Are those sensors sucking in infrared or visible light or both? If it has a radar onboard, what frequencies does it operate on? What's the resolution of every blasted electro-optical and radar sensor on that spacecraft? What can VenezSat see, hear, and *do*?"

His hard eyes swept the room, boring into those of his suddenly very quiet staff. Sawyer was generally easygoing, but he also had a knack for slicing through nonsense like a saber through warm chocolate. "People," he continued, speaking so softly the colonel in California strained to hear every word emanating from a tabletop videoconference speaker, "I want to know *intent*. And I want to know it after one or two revs, not six months from now. Look, Chavez may be a sick blowhard who gets a kick out of thumbing his nose at the United States, just to get his acne-pocked mug on TV and the front page every few days. But the little bugger is also dumb like a fox, so let's not write off his smallsat as simply a 're-search' bird just yet. Copy?"

"Jay, get your space-intel team working on this VenezSat, and co-ordinate with our intel shop here," Sawyer added, raising his voice and addressing the colonel's image beamed from 14th Air Force's

JSpOC. "We've spent a ton of money upgrading our 'space situational awareness' capabilities, since General Chilton sat in this chair four years ago. Now, people we answer to at STRATCOM and the Pentagon expect to see some solid returns on those SSA investments. So, what's the payload on that damned VenezSat? Is this a single platform or are there 'riders' on it? If there *are* parasites, why are they hitching a ride and what are those riders' capabilities? Is this satellite doing what we'd expect a research bird to do? Is it maneuvering, and, if it is, how far can it go and what can it do? Here's the bottom line, folks: Is that little 'research' bird a threat to America's space assets? If it is, what the hell can we do about it?"

The general leaned back and ordered the colonel to wrap up his briefing. As Jay did so, Sawyer took perverse delight in seeing his uniformed and civilian staff furiously scribbling in notebooks and thumb-typing on handheld communicator-data assistants. *We'd better come up with some solid answers*, he thought, jaws tightening. He couldn't explain why, but Sawyer had a bad feeling about Chavez's new plaything in orbit. As long as Venezuela's dictator was mouthing off about giving money to poor people, making a big deal about sending oil and food to North Korea, or nationalizing yet another key domestic industry, Sawyer could ignore the little twerp. But Chavez now had the ability to look down on America's heartland, maybe several times a week. The crafty Venezuelan socialist had stepped onto Sawyer's turf. As the nation's chief space warrior, Buzz Sawyer couldn't ignore that.

STARFIRE OPTICAL RANGE/KIRTLAND AIR FORCE BASE, NEW MEXICO

Seated at a sophisticated, multiscreen computer workstation, Air Force Captain Sandra Kindle cross-checked data from the Space-Based Space Surveillance system, verifying that twin SBSS satellites were, indeed, tracking and documenting Venezuela's new spacecraft. A second oversized monitor displayed a crisp, visible-light image of VenezSat acquired by the Starfire Optical Range's powerful telescope. With its 11.5-foot-diameter primary mirror, the telescope could capture images of basketball-size objects orbiting 1,000 miles above the Earth.

Intriguing configuration, she mused, studying the image with a practiced eye. Small solar panels, a few shadowed ovals—probably optical ports—and an S-band antenna were about all she could make out. But the satellite body's unusually dark skin and faceted shape were unlike any the young space operations and test engineer had seen before. The coloring and shape seemed to deflect light somehow, complicating Kindle's task: characterize VenezSat and report the findings to 14th Air Force's space-intel office.

"Okay, I have a solid lock-on, Morad," she called. "Ready to paint the target." Dr. Morad Romero, a bearded, heavyset man, appeared beside her, bending to peer closely at Kindle's workstation screens. He breathed heavily.

"Hmmmph. That as good as it gets?" he growled in a deep, surprisingly melodic bass. "Can't you drive in tighter?" Kindle nodded, zooming until the image filled a 30-inch screen and the satellite's sharp edges started to pixilate, then retreated slightly.

"That's the best we'll get in the visible-light mode," she said, looking up at the civilian contractor. Romero was a mountain of a man, but not overweight. His barrel chest tapered to a thick waist and hips, and he moved with a smooth, fluid ease, a testament to the excellent athlete he had once been. A mass of unruly black hair, thick eyebrows, and a profuse, gray-flecked beard hid most of the man's skull and features, but dark, flashing eyes immediately drew one's attention. They bespoke a razor-sharp intellect and focused intensity that matched his rapid, yet efficient motions. He had to be at least fifty years old, Kindle estimated, but he could pass for much younger.

"Want to see it in the infrared?" she asked.

"Naw. We already checked the IR scans. Didn't show much," he rumbled. Romero straightened and scratched his hairy chin, deep in thought. Kindle admired the hulking man, and, in an odd way, found him attractive. She was barely thirty years old, divorced, and pretty damned good-looking, she knew. Being older, Romero was hardly the type she'd normally give a second look, but working with him ten to twelve hours a day behind cipher-lock doors in the windowless Starfire labs had revealed intriguing facets of the man. After receiving a doctorate from the University of Arizona in 1980, he'd worked continuously for Ball Aerospace & Technologies in Boulder, Colorado, becoming one of America's best-known scientists, specializing in the applications of lasers in space. Today, that expertise would be put to the test.

"The bird obviously has several low-observable features," Romero mused softly, more to himself than Kindle. "Faceted surfaces, angled-back solar arrays, tightly hooded optics. All designed to deflect incoming energy. Fascinating . . ."

"Sort of like the B-2 bomber's stealth design?" Kindle asked. She, too, had specialized in laser systems as an undergraduate optics-engineering student. "Stealth" aircraft, such as the B-2 Spirit bomber, she'd learned, were designed to minimize detection by radar's radio-frequency pulses, but were less effective at hiding from modern lasers. Spacecraft rarely exhibited stealth features, though.

"Yep. Bounce those radar pulses away, not back to the transmitter. Just like the B-2," Romero agreed. "Looks like our Russian friends built VenezSat to hide from ground-based radar, all right. Now, why would a research satellite need to avoid tracking radars?" he asked rhetorically.

Kindle flashed her mentor an admiring, engaging smile and suggested, "S'pose Mr. Chavez has something to hide?"

Romero chuckled, a deep-throated, pleasant sound. "He just might, my dear. I *say*; let's take a peek under the bonnet of Mr. Chavez's sneaky package? What?" the big man said, affecting a mock English accent and shooting Kindle a mischievous wink, dark eyes dancing. *That* playful, conspiratorial attitude, plus Romero's ultrasharp intellect, were what attracted Kindle. She awkwardly scrunched her smooth, olive-toned features, trying unsuccessfully to return the wink. Failing, she resorted to an enthusiastic thumbs-up agreement.

As co-workers, the two had grown close in recent months, precisely because they shared very bright intellects, plus a somewhat perverse delight in using their technical talents to bend the rules of what Romero called "official dumb."

Romero shoved a fist skyward and silently mouthed an enthusiastic "Yes!" Seconds later, he hunched over a keyboard at an adjacent workstation, preparing Starfire's stable of lasers for action. Kindle switched one of her screens to display a long stream of alphanumerics, cryptic lines of letters and numbers that depicted the laser system's state of readiness.

"Power levels and temps are in the green," she clipped, all professional. "Portals double-checked; full open. Guidestar ready to fire."

Romero glanced across at Kindle's second screen, confirmed that VenezSat was still being optically tracked properly, then tapped an "Enter" key at his workstation. "Fire one!" he called in hushed tones. Kindle smiled, loving the conspiratorial demeanor the Ball Aerospace scientist adopted at times like this.

She responded with a staccato of Guidestar data readouts, giving Romero information he needed to ensure the main "frequency-agile" laser would work its magic when fired. She envisioned the faint, blue-green Guidestar laser beam racing into the clear sky above Starfire's observatory, a complex of low buildings tucked into a remote corner of Kirtland Air Force Base. The thin, clear New Mexico air at 6,240 feet above sea level barely registered that beam slicing through the atmosphere, bouncing off sodium atoms along its path. Light reflected from the atoms was detected at Starfire's ground site, providing critical information about turbulence, varying air density, and wind shears high in the atmosphere. Kindle confirmed the data had been accepted by a computer that, in turn, commanded dozens of tiny actuators affixed to a flexible mirror and series of lenses.

"Adaptive optics are working their magic, Morad," she said, eyes flicking between her workstation's dual screens. Data confirmed the lenses and mirror in the primary laser system—its "adaptive optics" train—were constantly being reshaped, compensating for atmospheric aberrations sensed by the pathfinding Guidestar laser. Those shifts would continually adjust the primary laser's properties, ensuring the beam literally held together as it passed through the air, en route to space and its VenezSat target. But Romero wasn't quite ready to fire "Big Light," as he called it. He grunted, a response Kindle recognized as a sign of focused, total concentration.

The big man muttered something unintelligible as his fingers raced across the keyboard, ending with a flourish and dramatic pinky-strike on the "Enter" key. Instantly, the primary frequency-agile laser fired through Starfire's now-open, protective dome, boring through the air at 186,000 miles per second, its properties being altered thousands of times a second to preserve the beam's structural integrity. The narrow thread of coherent light followed the lead of that previously fired pathfinding Guidestar laser, until it played across the features of Hugo Chavez's orbiting research satellite.

The beam's multiple visible-light, plus near- and far-infrared wavelengths became angstrom-size fingers that probed and searched every nook, port, and crevice on VenezSat's exterior surfaces. In an instant, laser light reflected from the spacecraft's myriad features bounced back to Earth and was captured by Starfire's array of extremely fine-tuned sensors. Powerful image-processing computers deciphered subtle messages embedded in the reflected laser light, then displayed the information on a screen Romero's eyes scanned intently. He finally leaned back in his chair, smiled broadly, and turned to Kindle, raising a palm.

"Gotcha, ya little bugger!" he stage-whispered. Kindle leaned to her left and smacked the man's open palm in a high-five salute. *God, how corny! And oh-so nineteen-eighties*, she thought, smiling broadly. A high-five was Romero's embarrassingly dated, engineer-nerd equivalent of a post-touchdown dance in a football field's end zone.

"Look at this, Sandy!" he enthused, sliding his chair aside. Kindle rolled hers over to Morad's workstation and searched the oversized monitor.

"Sooo . . . what am I supposed to see?" she asked, puzzled.

"We lucked out!" he enthused. "Some of Big Light's multiple wavelengths literally resonated with the satellite's optics, which allowed the system to characterize the entire optical train. See these?" He pointed to a string of numbers, then pulled the wireless keyboard closer and tapped a burst of obtuse code. The tabulation of numbers disappeared, replaced by a steadily growing, ever more-detailed wire drawing.

"That *is* the whole optics system! Holy . . . !" Kindle breathed, impressed. Romero's rumbling bass chuckle answered.

"Like I said, lucky. Doesn't always work like this." He leaned closer to the screen and typed another series of commands, causing the see-through wire-like image to grow in successive jerks, as if zooming in closer. "Hmmmm. I'll be damned . . ." he said, frowning.

She waited, but Romero only stared, thinking. "Well? What, Morad? Come *on!*" she insisted. The guy could be *so* damned exasperating!

He nodded absently. "I think we've got a stinker here, Sandy," he finally said. The earlier effusiveness was gone, replaced by a dark, sober scowl. He pointed at a tiny, tube-like object on the screen. "See that? Two bits says it's a dad-gummed laser. And those little circles? Some kind of separate, stand-alone optical train. Now, what the hell . . . ?"

Kindle glanced at the man and raised an eyebrow in surprise. Romero rarely cursed. Privately, she'd told friends the guy "couldn't say it, even if he had a mouthful of it!" Obviously, he was worried, which alarmed her. Morad *never* worried!

Grasping a trackball, she took control of the workstation, expertly manipulating the wireframe image, as had become their unspoken custom. Morad performed the digital-code magic and Kindle handled the graphics. A generational thing, she assumed. Zooming closer, she bracketed the suspected mini-laser and its linked series of what appeared to be lenses, employing a series of click-drags via the trackball, then tapped a few keys. Momentarily, a box jammed with numeric data appeared on the screen, the product of a sophisticated analysis algorithm they'd written. Both studied the data in silence.

"Morad, I'll see your two bits and raise you a buck," she declared. "Buck-twenty-five says that's a laser-ranging system."

Romero glanced sideways, shooting an admiring half-grin. "Hey! Gettin' good, young lady! How'd you pick that up so fast?" Indeed, once she pointed out a few telltale components, he, too, recognized the undisputable characteristics of a compact laser-ranging system. But not before. His protégé was coming along very nicely!

The two laser specialists discussed a few technical points and agreed they'd nailed the critical features of VenezSat. But they disagreed about how to proceed. Romero argued for running a few more confirmation checks, whereas Kindle advocated getting their first-cut info to 14th Air Force immediately. Kindle was still toying with the wire diagram on the screen, studying the image closely. Romero glanced past her to check one of Kindle's workstation monitors, did a quick double take, and snapped upright.

"Oh, shit! Look!" he breathed, knuckle-tapping Kindle's upper arm. She quickly rolled the chair back to her own workstation. Imagery beamed from the Maui Optical Tracking Identification Facility telescope showed VenezSat had rotated from its original position. It now stared directly at the ground-based telescope, as if an alien intelligence were onboard.

"That's not good, is it?"

"Definitely not. The damned thing knows we laser-dazzled it!" Romero said, surprised. Kindle's fingers immediately went to work, typing a familiar series of commands, ordering the computer to store and back up billions of data bits. In seconds, thousands of images and their associated numerical data had been captured, available for future assessment, if necessary. Somehow, she knew considerable scrutiny of their imagery *would* become necessary, and very soon.

Romero pointed to a box on the screen, a box linked by a thin green line to the image of a sunlit Venezuelan spacecraft staring at him. Several of those digits had started changing, as if they were on a rolling tumbler.

"It's moving!" Kindle exclaimed. She, too, stared at the unnerving image, which stared back. It seemed to be growing, slowly but inexorably. "Damn, Morad! That SOB's coming after us!"

He laughed, the rich bass breaking a tension that had gripped the engineers. "No, I don't think so, my dear. But it's definitely moving."

"Damn, did *we* cause that, Morad?" Kindle asked, eyebrows knitted in worry. He was struck by how young she suddenly seemed. A terrific, talented engineer and seasoned Air Force officer, Captain Sandra Kindle rarely exhibited even a hint of stereotypical gender vulnerability. To the contrary, she radiated a strong, don't-mess-with-me aura that tended to intimidate would-be suitors. But that tough armor had chinks. Her almost-fear was vaguely unsettling, forcing

Romero to hesitate. He looked at the big screen, trying to summon a smart, reassuring answer. He drew a mental blank.

"I suspect we did—but inadvertently, Sandy," he finally sighed, choosing his words carefully. "I suspect our Ruskie friends installed an automatic RF/laser detection system on VenezSat, similar to what we're flying on our own national security spacecraft these days."

"Radio-frequency and laser sensors," she nodded. "Sure. Makes sense." Laser and RF detectors had become military and intelligence satellites' first-line sentries in recent times, designed to automatically sense when their hosts were being "dazzled" by lasers or "painted" by radar or some other form of electromagnetic energy. These detectors had become standard equipment on government-owned spacecraft in recent years, thanks to Chinese probing of on-orbit U.S. and allied platforms in the early and mid-2000s. Linked to other onboard systems, these tiny electronic sentinels could automatically trigger protective measures, such as slamming covers shut to shield exposed optics from laser energy, and "auto-safing" circuits that guarded electronic systems from damage. Unfortunately, only a handful of newer U.S. satellites were so equipped, leaving aging legacy spacecraft vulnerable to interference, whether intentional or not.

"What do we do now? Are we in deep kimchi?" Kindle asked.

"Well . . . I doubt it. You found the target satellite and captured all relevant data. We were in the process of 'characterizing' VenezSat, which is exactly what Space Command ordered, right?" Romero mused. She nodded, but kept nervously eyeing that still-staring satellite image. "And I suspect we're probably the first to detect a change in VenezSat's orbit and activity. So, as soon as we call the Jay-Spock at Fourteenth, I'd bet money we'll be hailed as frappin' heroes," he said, striving for an upbeat air.

Kindle grimaced and shot Romero a blazing glance. She recognized B.S., especially when it was flowing profusely. Nevertheless, she nodded in agreement. "Right. Okay, if you'll assemble all the photos and data into a compressed file, I'll bang out a summary report," she offered. Again, it was a division of labor they'd worked out long before. Morad was far more skilled at manipulating data and imagery than composing well-structured reports laced with military jargon. Kindle was definitely a more-talented govern-mentese wordsmith.

"Any idea where VenezSat is headed?" she asked, again watching Romero closely. He shook his tousled head vigorously.

"Nope. But once we get this data to the First Space Squadron toads, they'll work their trajectory- and conjuctive-analysis magic on it." He hesitated, shooting her a dark glance. "I don't have a good feeling about this thing, Sandy."

STRATCOM/COMMANDER'S OFFICE

General Howard Aster's lanky frame was hunched over a computer keyboard as the STRATCOM commander sorted through a multi-tude of classified e-mails.

During a special update briefing for Major General Dawn Erikson and her NSSI team in the wargaming center, Aster's handheld com-municator had fielded a red-bordered text-message alert from Buzz Sawyer. Buzz was forwarding a critical report from Air Force Re-search Lab's Starfire site in New Mexico via a classified network. The Space Command four-star urged his STRATCOM counterpart to "take a look ASAP."

Aster finally located the file, opened it, and skimmed the curt, professional report. He turned to the attached images, studying each stark, black-and-white photo of Chavez's VenezSat for a few moments. Swearing through clenched teeth, he forwarded the re-port and images to Annie's laser printer.

He punched a speed-dial number on a desk phone, and verbally relayed the gist of Morad Romero's and Captain Sandra Kindle's re-port to Secretary of Defense Hurlburt's executive assistant, asking that the SecDef be given the information as soon as he returned from the White House. Aster swept through his outer office, wav-ing Burner Burns to join him. Burns had already retrieved the AFRL report and photos from Annie's laser printer.

"We've got more shit to deal with, Burner," Aster clipped, jerking a thumb at the documents Burns carried. The flight-suited aide struggled to keep up with his boss's long-legged stride, while scanning each page.

"VenezSat, sir?"

"Yep. That's Chavez's so-called 'research' satellite. The damned thing started maneuvering a few hours ago, and the AFRL experts at Starfire say it's apparently changing orbit," Aster replied, diving into a stairwell. Elevators were too slow to suit the general, Burns had learned.

"And . . . that's a problem, sir?" Burns asked. He was still trying to make the mental leap from flying B-2 bombers to understanding the finer points of satellites, orbitology, and other space-geek stuff.

"Could be. Maybe not, either. But Sawyer's space intel folks are concerned enough to wave the flag. Their space control team is tracking that bird, trying to figure out where it's going."

Aster rounded a corner, stopped abruptly near the wargaming center's closed double doors, and jabbed an index finger at one of the photos Burns carried. "The smart folks at Starfire say Chavez's new toy-in-space has a lot of interesting features, more than you'd expect of a research sat. I want General Erikson's NSSI team working on this one right away, 'cause I have a feeling we're going to get involved. Don't know how or why yet, but Buzz seems to think that Chavez could be up to something." The general retrieved the sheets and stepped through a side door of the center, searching the room.

Burns followed, noting that Colonel Jim Androsin, the center's chief, had allowed "controlled chaos" to resume, a fast-moving tactic to focus the energies of various wargaming experts on increasingly difficult "what-if" scenarios. Several knots of four-to-eight people huddled around workstations throughout the large, theater-like room. Uniforms from four U.S. services mixed with those of British, Australian, and other allied officers. A fair number of civilians embedded with these groups had abandoned their coats and ties long ago. A strong odor of coffee laced the stale air.

A handful of key wargamers had been in the windowless center every day, dealing with the fallout of what news reporters were calling "America's 9/11 in Space." Losing dozens of national security, civil, and commercial satellites—to say nothing of five astronauts on the International Space Station—last year to attacks by that drug cartel–funded maser fired from Dushanbe, Tajikistan, had shaken America's space community, political leaders, and citizens to their boots. A national sense of crisis had been exacerbated by a nuclear-tipped missile fired from Iran; North Korea's nuke weapon detonation at the fringes of space, causing the demise of many more spacecraft, and, finally, the stunning double-pronged terrorist attacks in Southern California.

STRATCOM's wargamers were trying mightily to think beyond today, exploring a dizzying array of "alternate futures" that the nation and its allies might face on multiple fronts. Their insights had proven invaluable already, and both the Pentagon and White House were looking west, to people in this room at Offutt Air Force Base, Nebraska, for critical updates and recommendations. That spotlight from Washington had ratcheted up both the interpersonal tension and a grim sense of determination throughout the center, Burns observed.

Colonel Androsin and his BOYD system information-technology wizard, Jill Bock, were still in the center's "Pit" at the front of the amphitheater. Bock was seated at her workstation, surrounded by Androsin, Major General Dawn "Viking" Erikson and two of her NSSI experts, both lieutenant colonels wearing tan-and-green battle-dress camouflage. Erikson's aide, Major Hans Richter, was seated, while the others stood, clustered behind Bock.

Aster approached and laid an arm across Erikson's shoulders, drawing her attention. From across the room, Burns noted the blond officer's features brighten and a broad smile appear as she turned and looked up at Aster. The two generals stepped aside, and were soon engrossed in perusing the VenezSat report and photos.

"Because it might have a ranging laser on it, General Sawyer thinks this bird poses a threat?" Erikson summarized, studying the photos.

"Roger that. Until it started moving, nobody considered it a problem. But Buzz says, 'any satellite that can maneuver is a potential threat.' His folks are running conjunctive analyses, trying to determine where it's going," Aster said, stifling an escaped yawn. He looked across the room, catching Burns's eye. The general tipped a cupped hand to his mouth.

More coffee. The boss is smokin' through more brew every day, Burns thought, heading for a well-stocked coffee bar on the amphitheater's upper level.

This morning, the general's executive assistant, Annie, had confided that their STRATCOM boss was looking unusually tired—and so was Burns, she chided pointedly. A matronly, dedicated assistant for dozens of STRATCOM's previous commanders, Annie always worried about "her boys," and closely monitored their behavioral nuances. If you were slightly off, Annie noticed and was sure to inquire about your well-being.

Burns had admitted that his own butt was definitely dragging, a casualty of long days, short nights, and not a single day off in months. Like others around STRATCOM headquarters lately, he was undoubtedly getting a tad short-tempered and cranky, too. His wife had quipped that she was going to start making Lieutenant Colonel Burns show an ID card at the door when he finally *did* come home. Their two children had asked repeatedly, "Where did Daddy go? When's he coming home?" They thought he was on yet another Temporary Duty or TDY trip!

That one hurt, he reflected, filling a cup for Aster. Black, no sugar. He hesitated, then poured a second cup, in case General Erikson was also in need of a caffeine shot.

"If Fourteenth decides VenezSat *does* constitute a threat, what are our options, sir?" Erikson asked, as Burns offered her and Aster large, steaming cups. She flashed that perfect smile and mouthed a "thank you," accepting the dark brew and a couple of sugar and creamer packets from Burns.

Aster took a sip before answering, also nodding to Burns in thanks. "Not many, I'm afraid. Hell, you know the story. No 'Operationally Responsive Space' systems online yet. Yeah, we have one supersecret spaceplane—the Blackstar system I told you about—but it's down for a canopy replacement. Bottom line is, we have no means to put something in space on short notice. Absent that, I'd damn sure love to have a maser like the one that creamed our satellites last year. With a maser or some other directed-energy weapon, we could interdict an orbital threat on very short notice. Right now, we can't. We just track it and wait."

Erikson nodded, watching him over the edge of her cup as she blew across the liquid's surface, cooling it. Burns started to turn away and leave the generals to their private conversation, but Aster clapped his shoulder. "Burner, how about giving Speed Griffin a quick call and get a reading on Blackstar's status, would you? Tell him I might have to call up his super-bird again, and soon."

"Roger, sir. Will do."

Erikson donned a set of stylish, folding reading glasses and studied the VenezSat photos more closely. "Let's say Space Command decides this bird really *is* a serious threat," she mused. "To what, we can't say, but let's assume it is. Does STRATCOM have specific rules of engagement to deal with an on-orbit threat? I'm not aware of any, but . . ."

"None. We're making this crap up on the fly, Viking," Aster interrupted, staring into his coffee and swirling it. "That's why I need your and NSSI's insights to help us develop strategies we *could* employ, should VenezSat start acting up. You give me some options, and I'll have Androsin crank 'em into our wargame to assess the possible ramifications. Then we can act with a level of confidence, having considered a number of potential outcomes—*if* the need to act arises, of course. As it is now, we're not at all prepared for dealing with so many unknown unknowns."

"Understand, sir. We'll get on this. And I'll tap our space-law people back at the schoolhouse, too." She hesitated, cocked her head to one side, and eyed the tall four-star askance. "General, maybe I'm getting ahead of myself here, but would you consider taking . . . ummm . . . 'untoward' actions against that Venezuelan satellite?"

Aster shook his head and gave Erikson a knowing glance. "Nothing 'untoward,' Viking. But my job calls for looking at *all* options under our counterspace mission, right? Those may include 'space force applications,' so I want you to take a hard look at those possibilities, too. Let's leave it at that, at least for now. Okay?"

"Okay, sir. But, as you know, we're skating on uncharted ice here, and that ice will get very thin, at times. Based on our work at NSSI, some of the policy and doctrinal issues STRATCOM is facing are well above even your pay grade. I'd highly recommend that, before any additional counterspace or space force application missions are green-lighted, you advise the SecDef and other political leaders in Washington. We can't have General Aster starting a war in space on his own."

Aster nodded slowly, eyeing Erikson admiringly. *You are one knock-em-dead woman! And a hell of a gutsy officer! Tell it like it is, rank be damned,* he thought.

"Message received, loud and clear, Viking." He grinned broadly, underscored by a quick wink.

She returned the smile and ignored the wink, but held his gaze an extra-long beat. She felt her insides jump and a tingle of excitement. Whether that tingle was linked to a daring idea taking shape—a true "space warrior" option for dealing with VenezSat—or a woman's reaction to Howard Aster's warm, deep-blue eyes holding her own, she wasn't sure.

U.S. AIR FORCE SPACE COMMAND HEADQUARTERS

"Reconstitution! That's our top priority, people! We absolutely *have* to restore some of these critical space-based capabilities!" General Buzz Sawyer declared, jabbing a forefinger at a wall of plasma screens overlooking his command center. Five senior officers clustered around him, all dressed in olive-drab flight suits. The atmosphere was glum, at best.

Minutes earlier, the five officers had briefed Sawyer about the current availability of U.S. launch vehicles, and the bottom line wasn't pretty: Not a single Atlas V or Delta IV booster could be launched before mid-summer, at best. Those were the Air Force's primary workhorse spacelifters, and the only U.S. rockets capable of placing large, heavy satellites in orbit. A few small commercial companies claimed they could provide a vehicle to launch "smallsats" within the next few weeks, but Sawyer and his operations staff were skeptical.

"Sir, maybe we should tap Aerospace or Mitre to help us sort out these commercial-launch guys," suggested Colonel Greg "Gun-slinger" Gunsallus, a Space Command deputy for operations. The two firms often provided on-call technical support to the Air Force and other government agencies. In particular, Aerospace Corporation was the Air Force's go-to science and engineering experts for myriad space matters.

"Negative, 'Slinger," Sawyer replied, shaking his head. "Aero-space and Mitre are our primo Federally Funded Research and De-velopment Centers, especially when it comes to orbital mechanics and satellite-tech matters, but they have little or no interaction with the commercial sector. What we need is somebody who works both milspace and commercial space. There's gotta be somebody out there who not only understands the national security space business, but also speaks the language of space balance sheets, shareholders, and profits. Come on, guys. Let's get out of our brain-cubicles and *think*!"

An uncomfortable silence settled over the group. The officers glanced at each other, avoiding Sawyer's hard eyes. The four-star was becoming testier with each passing day, it seemed, and nobody wanted to risk winding up on Buzz Sawyer's shit list. Finally, Gun-slinger Gunsallus proffered an idea.

"Sir, the Space Foundation has its headquarters here in Colorado Springs, and it stays on top of both national security and commercial space issues. Could they give us some readings on small launcher companies?" he suggested, watching Sawyer carefully.

The commander crossed his arms and stared at Gunsallus. "They might, at that. Give 'em a call and see if one of their execs would mind coming over for a conversation." He slapped the colonel's arm and turned to leave the command center. "Good idea, 'Slinger. I want to talk to those Foundation guys ASAP."

Within twenty minutes, Gunsallus had secured a commitment from two of the Space Foundation's key executives; they'd be in Sawyer's office that afternoon. Both were retired military officers having years of space-related experience, and understood the co-nundrums Sawyer now faced.

Unless Air Force Space Command could restore dozens of lost

space capabilities, American warfighters would be relying on communications, intelligence-gathering, missile-defense, weather-forecasting, and navigation methods last employed during the Vietnam War era. Gunsallus was too young to remember those days and methods, but he knew they were "unsat" for today's ultra-lean, overstretched defense forces. Today, without space support, America's global war machine would soon grind to a crawl. And, if they sensed the world's only military leviathan was crawling around in the dark, a hell of a lot of bad guys would be poised to pile on.

AIR FORCE SPACE COMMAND/GENERAL SAWYER'S OFFICE

"You're convinced these four outfits absolutely, positively *can* get my satellites into orbit, and do it within the next few weeks?" Buzz Sawyer asked, gesturing at a wrap-up PowerPoint slide projected on the screen.

"I am. All four are limited to carrying fairly small payloads, and they'll have to work out interface issues between your satellites and their vehicles, but every one of these companies has the spacelift capability to get micro- or nanospacecraft into low-Earth orbit, and soon," assured Steve Trevino, a Space Foundation vice president and retired Army officer. He stood at the far end of a conference table in Sawyer's expansive office, now toying with a laser pointer.

"You concur with that assessment, E.T.?" Sawyer asked pointedly, half-turning to the man seated to his right.

"I do. It's important to note, General, that these companies are routinely flying so-called 'space tourists' on suborbital flights. They've all logged successful missions," E. T. Horn added. "However, two of them—SpaceX and Andrews Space—also have completed flights to LEO in preparation for the next phase of space tourism: letting customers spend a few days in orbit." The Space Foundation's president and CEO, Horn had retired as an Air Force colonel in 1999, then served several years as a corporate communications professional at Atlas Holdings before joining the Foundation.

His aggressive leadership of a small, handpicked team had rapidly expanded the Space Foundation's fledgling conferences and educational outreach programs. He also had made a solid impact in Washington, becoming an effective spokesman and proponent for the government, civil, and commercial space sectors. In recent years, the Foundation's influence also had embraced the rapidly growing international space community. Horn's team had ensured the Space Foundation carefully positioned itself as a nonpolitical space proponent.

"SpaceX has already flown . . . what? Two, three people to one of Bigelow Aerospace's habitats?" Sawyer asked, glancing back to Trevino.

"That's correct, sir," the vice president nodded. "SpaceX's Falcon-Nine rocket is now man-rated, and capable of reaching LEO with a respectable payload. It flew Bigelow's first three astronauts to the company's new Sundancer inflatable module early this year." Trevino glanced at Horn, who took the handoff cue with practiced ease.

"General, this is proprietary information, but I cleared it with the company before coming over here. In fact, Bigelow Aerospace *wants* you to know about this, because their new CEO thinks you might be able to help him . . . and he'll gladly return the favor at some point. Bigelow's been working closely with SpaceX this past week to launch another Falcon-Nine ASAP. They're desperate to get those three astronauts off Sundancer, and SpaceX is . . ."

"Because of that damned North Korean nuke?" Sawyer interjected.

"Exactly. Sundancer is a bare-bones outpost, designed only to test and mature a number of systems, so they don't have sensitive radiation detectors onboard yet," Horn explained. "But Bigelow's astronauts are scared stiff, wondering if they've soaked up life-threatening doses from that nuke."

Through clenched teeth, Sawyer muttered an obscenity and colorful comment about the North's bad-haired chairman. Standing to indicate the meeting was drawing to a close, he thanked Horn and Trevino for sharing their insights, and assured them that his command would do whatever it could to help Bigelow and SpaceX get those astronauts back on Earth safely.

At the door, Sawyer shook hands with the Space Foundation executives and asked, "Would you mind giving those spacelift companies a heads-up that my people will be contacting them? We absolutely *have* to reconstitute some critical space capabilities as soon as possible. When our first payloads are ready, we'll be looking for a ride to orbit. If your companies can deliver, and do it on time, this could be the beginning of a whole new model for national security spacelift."

9 THE UNTHINKABLE

MOUNTAINS OF WESTERN IRAN

Hassan Rafjani closely watched the final stages of fueling, silently marveling at the tall Shahab-4 missile's bulk as it seemed to awaken, soon to become a sinister, fire-breathing dragon of death. The hiss of venting liquid oxygen, accompanied by clouds of the gas swirling around silver-suited technicians, gave life to the ominous, olive-green rocket.

Dagger's eyes were drawn repeatedly to the missile's nose cone far above. It housed the only remaining nuclear warhead Iran possessed, and Rafjani was acutely aware that years would pass before his beloved Persia could build another of the fifteen-kiloton atomic devices. Months ago, an infidel attack on Parchin, Iran's primary uranium-refining site, had destroyed thousands of precious centrifuges and killed dozens of scientists, engineers, and skilled technicians, virtually eliminating the nation's uranium-enrichment program in an instant of white-hot explosion and fire. Only a few thousand of the sophisticated centrifuges remained, buried deep under the sand and rock of Natanz, but they were insufficient to fuel the urgent needs of Iran's secret nuclear weapons program.

How the Americans, Europeans, or Jews had destroyed Parchin—

particularly without leaving a trace of explosive material—continued to baffle him and his few remaining nuclear scientists. Unquestionably, though, it had left Iran's prized "Islamic Bomb" effort in shambles. Only this single warhead had survived the Parchin attack, thanks to his, Rafjani's, foresight. Two days before the mysterious explosions, he had ordered the nation's last operational warhead removed from storage at Parchin and trucked to this tunnel deep in the Zagros Mountains of western Iran.

Subsequently, the warhead was mated to its Shahab-4 steed, ready to launch within hours of a "go!" order. Soon, he would give that order and unleash the final, beautiful, cleansing via Allah's Sword. That singular act would ultimately fulfill the prophecies, enabling the Mahdi's blessed return, God willing. Soon, the Mahdi—Islam's revered Twelfth Imam—would lead the world's oppressed Muslims into a golden age, an era free of infidel Christians and Jews; free of the West's sinful music, movies, television shows, and other trash. But first, he, Hassan Rafjani, Allah's most-loyal servant, must cleanse the Earth of Jewish and Western vermin.

"Have you double-checked the Russian Glonass receiver's operation? And verified the correct target coordinates are loaded?" Rafjani asked in hoarse tones, more a demand than a query. With considerable difficulty, he had acquired a missile-guidance system that relied on Russia's Glonass navigational satellite system, but he remained wary of the unit's reliability. After all, a bootlegged American GPS guidance system had failed, causing a nuclear-armed Shahab-4 missile to fall harmlessly into the sea. His guidance engineers had been unable to determine exactly why that missile's guidance system and the weapon's detonator circuits had failed so miserably.

Subsequently, the European Galileo receiver he'd purchased from a French company had proven to be worthless, after two of the nascent Galileo constellation's three satellites had died unexpectedly in orbit. Rafjani still seethed at those twin failures, which had embarrassed him before the fainthearted mullahs in Tehran. He would *not* fail again.

"I personally have verified the satellite navigation system is fully operational, sir," a trusted Quds rocket engineer assured. "And the correct target coordinates have been loaded and validated. Soon, the holy Shahab will be ready to launch."

Rafjani again surveyed the missile, which had been towed from its protective tunnel, then raised to a vertical position atop a transporter-erector system. Bathed in bright xenon light, the rocket seemed regal, poised, and anxious to fulfill the prophecies. The blue-eyed Rafjani felt a cool calmness sweep through him, as he stared at the thick-bodied missile. In this powerful weapon, he literally could feel the Mahdi's presence and hear his ethereal voice whispering in Rafjani's mind. *Insha' Allah!*

"It must launch before dawn, understand?" he ordered. The engineer nodded confidently and departed, leaving Rafjani alone. Whispering, he prayed to Allah, concluding with a promise: "This day, the Jew-serpent's head will be severed. Never again will its poisonous bite harm a Persian, Arab, or Palestinian."

The radio in his hand came to life, ordering all personnel into protective bunkers. The hour for launching the final phase of Allah's Sword had come.

NORTH AMERICAN AEROSPACE DEFENSE COMMAND UNITS/COLORADO

A staring, heat-sensitive sensor onboard a Space-Based Infrared System satellite in geosynchronous orbit, high above the Atlantic Ocean, first detected the Shahab-4's fiery plume as the rocket streaked skyward from western Iran. The compact streak of flame triggered a missile-defense warning alarm, immediately drawing the attention of an Air Force controller at Buckley AFB, east of Denver. Shortly thereafter, a series of automated alerts sounded similar alarms at the North American Aerospace Defense Command (NORAD) operations center. Recently moved from the granite depths of nearby Cheyenne Mountain, the center was now housed deep within a windowless complex at Peterson AFB on the edge of Colorado Springs, about 80 miles south of the Buckley AFB ground station.

The SBIRS staring sensor locked onto the Shahab's white-hot

tail, tracking the rocket's ascent and beaming a steady stream of digital data to an array of ground stations. In the NORAD Missile Defense center, a massive, wall-size display screen depicted the Shahab's projected course and, soon, a large, red oval that overlayed a map of the Mediterranean region. That oval represented the missile's potential target area.

A NORAD command director initiated a standard missile-event conference call, and began a litany he and countless predecessors had rehearsed daily for decades. Because NORAD was a binational air and space defense organization, the Prime Minister of Canada quickly joined the missile-warning teleconference, followed by Pierce Rutledge Boyer, President of the United States.

Within seconds, the watch team supplied critical data that enabled a rapid, decisive conclusion: "North America is not, I repeat, *not* under attack," the bird-colonel command director assured. "Missile is an Iranian Shahab-Four; warhead type is unknown, but assumed to be nuclear. Warhead impact zone is projected to be Israel or the eastern Mediterranean. Data are being relayed to missile defense batteries in Israel and U.S. naval forces in the Med. Satellite communication links are erratic. . . ."

"My God," Boyer breathed. "That damn fool Ahmadinejad did it again! But *why?*" The teleconference link remained silent. Every listening participant had the same question, but was focused on updates from NORAD, waiting . . .

The command director's professional, measured dialogue continued, "We have separation. Warhead still on projected trajectory; no indication of maneuvering. Rocket body is reentering." NORAD's big screen depicted that red-colored, projected-target ellipse shrinking, until it was completely inside a map of Israel.

"Update to track," the command director clipped, his tone reflecting the strain of new knowledge. "Projected impact zone is Tel Aviv, Israel. Repeat, impact zone includes Tel Aviv."

"Time to impact?" the Canadian prime minister asked sharply.

"Estimate one minute fifty-eight seconds to impact. Solid infrared track. Arrow and Patriot missile-defense batteries confirm receipt of infrared-track data. All interceptor units cued. No radar lock-on, as of yet." Other satellites, such as America's aging Defense Support Program missile-detection birds orbiting about 22,000 miles above the equator, also had picked up the Shahab's hot rocket plume, and were providing infrared-tracking information to complement that of the newer SBIRS spacecraft.

As the Iranian warhead arced over at the fringes of space, high above Iraq, then dipped earthward above Jordan, its nose aimed at Israel, long-range radar systems in England, southern Europe, and Israel detected, then tracked, both the flamed-out Shahab missile body and a deadly warhead. Those radar data provided vital "dual phenomenology" confirmation. With both space-based infrared and ground-based radar systems cross-checking track data, there could be no doubt: Iran had launched a long-range ballistic missile and warhead at the national capital of Israel.

UNDERGROUND MISSILE DEFENSE CONTROL
FACILITY/ISRAEL

"How much time do we have?" asked Moishe Baron, Israel's prime minister.

"Slightly more than a minute, sir," an Israel Defense Forces general replied, clutching a red telephone receiver tightly. His tone was professional, but the heart behind the voice was racing, threatening to tear itself from his chest. The officer's wife and two children were in Tel Aviv.

"Then our lives are in your hands, General," Baron said softly. "Keep this line open . . . and Hashem be with you and your forces." The general knew Baron, too, was in Tel Aviv. Should the phone line suddenly go silent . . .

Deep in a bunker beneath the Negev Desert, the general's team of highly trained missile-defense professionals was a blur of practiced, fervent activity. Clipped orders were implemented in seconds, preparing Arrow air defense interceptor missile batteries to fire. Mistakes were not an option; there would be no time for a second round of launches. Either the Arrows destroyed that Iranian warhead, or Tel Aviv would be evaporated.

A ring of Patriot air-defense batteries on the outskirts of the city would have a second shot at the warhead, should the general's Arrows miss their target. But a close-in Patriot engagement meant debris would fall on the Jewish homeland, exposing citizens to potentially deadly radiation. That is, *if* the Patriots could handle the demands of a supersonic-maneuvering end-game intercept. The Arrows *had* to succeed.

"*Ferkocht!*"

A curse in Yiddish drew the senior officer's attention. "What's the problem?" he barked.

"Sir, American track-cueing data are dropping out. Not getting solid updates as fast as we should!" a frantic operator replied.

"Downlink signals are intermittent!" yelled another sergeant, a uniformed woman with dark hair tied in a tight bun at her neck. "Comsat data links between the U.S. and Israel are severely degraded, sir. Some birds are completely inoperative; others are damaged. That's why American SBIRS and DSP satellite information have dropouts!"

The general's stomach contracted again. Without solid track data, his Arrow interceptors would be shooting in the blind, literally. "Cue the Arrows to the American infrared data, but hold until you have solid radar-track information," the officer ordered, perhaps too loudly. There were only six people crammed into the small concrete bunker, which had become stiflingly warm. Heat from computer equipment, radios, and power supplies was accentuated by anxious, rapidly breathing operators. The room's tension was palpable. He could smell, as well as feel the stress.

"General! Updates, please," the prime minister's calm but firm voice demanded.

"Sir, we're standing by for radar-track data," the officer replied. *Where* is *the damned radar data?* he wanted to scream. Then he saw it appear on an operator's computer screen. First from Fylingdales in England, then a site in northern Israel, radar data streamed across the screen.

"Radar data's dropping out, as well, sir!" the same operator announced in alarm. She tapped a keyboard frantically, trying to coax a contiguous graphical track from the machine. A hazy band appeared, not the solid, unwavering arc the operator so desperately wanted.

"Dump everything—all the data you've got—into the track-fusion system!" the general ordered. He was sweating, breathing in shallow pants. The Shahab-4's warhead was streaking downhill now, approaching Israel at almost four times the speed of sound. "Get a track on that damned thing! I need a targeting solution! *Now!*"

His operators worked with practiced speed and purpose, but they *had* been on the verge of panic, the general realized. His harsh order had yanked each of them back to their duties, personal concerns shoved to the background, enabling practiced, ingrained procedures to take over.

The general forced himself to breathe deeply. *Get a grip!* he commanded himself. A decades-old image flew across his mental screen, a flash of silver-wing leading edges spitting bursts of tiny red sparks. Those had come from Egyptian MiG fighters, enemies shooting at *him* and his F-4 Phantom years ago. He recalled the almost uncontrolled panic that had seized him then, fear tempered only by the stark knowledge that succumbing to it would mean certain death.

Now, as then, he choked down the mind-scrambling, throat-seizing fear by forcing himself to relax, to think, to rely on a gut-level confidence that he already *knew* what had to be done. Panic always translated to screwing up and, at worst, dying. But it also made one look bad in front of subordinates and superiors. Unacceptable. He was a battle-honed warrior, and true warriors did *not* screw up. Warriors performed well under fire, *exceptionally* well.

His mind cleared. The brain-static of fear and runaway "what-ifs" washed away, replaced by a vaguely familiar clarity and calm that he rarely experienced in peacetime. In a way, he was back in the F-4's cockpit, totally immersed in the present, an omniscient being centered in the "now" of combat. First answers, then decisions, finally came.

"Stand by to fire Arrow Battery One," he called. "Two missiles, ripple fire . . . *Ready* . . . !" The Shahab's warhead track had solidified before his eyes, thanks to a temporarily stable satellite communication link . . . or the Lord's hand. Maybe both. Either way, the general instantly decided, then acted.

"Fire One!—Fire Two!" he clipped, commanding two Arrow

interceptors into the air. Separated by mere seconds, the twin solid-fuel rockets raced into the gray-blue of Israel's desert sky, not slowly, as one would expect of a satellite-launch vehicle. But blindingly fast, rocket-boosted racehorses leaping from the gate. Two dense smoke trails traced high-speed paths as the missiles streaked to an imaginary spot in the dark purple of the upper atmosphere. That spot's location was continually updated by a stream of digital data beamed from a ground antenna. In turn, onboard computers commanded small aerodynamic-control vanes on each missile's aft body, ordering them to twitch, altering the interceptors' flight tracks. The Arrows quickly reached supersonic speeds, but kept accelerating, as their warheads automatically armed, ready for space combat.

"Arrow One, ten seconds to intercept; Arrow Two, twelve seconds," an operator called in a rapid staccato. "On track."

Eyes glued to their own monitors, yet flicking to a large display on the control bunker's wall, each uniformed operator silently willed the two Arrows to a spot in space that would soon be occupied by that deadly Iranian warhead. Minds were time-sharing, though, uttering silent prayers in parallel with the performance of essential duties.

"Mr. Prime Minister, two Arrows are guiding properly, less than ten seconds from intercept; on track," the general said, pressing the red receiver hard against an ear. Moishe Baron didn't reply, per standard missile-defense procedures. His comments were unnecessary and irrelevant, at the moment. Those of the intercept officer were critical.

Ten seconds. They could be my last, Baron thought. The flash of insight was a bolt of stunning clarity, truth uncluttered by qualifications or "maybes" or "what-ifs." Either the Arrows intercepted that Iranian warhead, or he and millions of Jews died.

On the intercept-control bunker's big screen, a crimson arc representing the threat missile's deadly path merged with the twin, slightly offset green arcs of two Arrows.

"*Impact!* We have an intercept!" an operator shouted. The room froze briefly, then erupted in deafening noise. The first Arrow's green arc had disappeared, as had the warhead's red one. The second Arrow's track continued through the merge-icon.

In the rarified air of Earth's upper atmosphere, the first Arrow interceptor's proximity fuze had sensed the presence of a supersonic Iranian warhead, waited briefly, then detonated. White-hot fragments spewed in a broad fan as a focused, cone-shaped pattern of steel tore into Iran's one and only nuclear warhead, shredding it. A flash of light marked a secondary explosion, as the warhead's carefully machined wedges of high explosive (HE) ignited.

But there was no perfectly timed detonation of those wedges, no coordinated explosions that would have compressed a tiny ball of fissile material at the bomb's core, triggering a supercritical chain reaction of splitting atoms. Instead of triggering a nuclear detonation, the few segments of HE exploded, tearing the warhead's nuclear "primary" apart and scattering Iran's precious uranium. The deadly warhead had been neutered, its fragments falling in high-speed, smoky arcs, mere radioactive debris.

"Mr. Prime Minister, we've confirmed the first Arrow has destroyed the threat warhead," the general breathed, soul-level relief apparent.

"The Lord's mercy!" Baron sighed, barely audible. "Again, He has parted the Red Sea."

The phone line was silent for a long moment, as each man struggled to grasp the magnitude of what had just transpired . . . as well as what had been averted. The blinding speed of intermediate-range ballistic missiles and supersonic interceptors engaging at the edge of space was difficult to envision. Such an intercept was akin to hitting a bullet in-flight with another bullet—a near-impossibility for a human, but an amazing, high-probability encounter made possible by modern computer-driven systems.

"General, please convey my sincere appreciation to your team. . . . And that of all Israelis, too," Baron added, a catch in his somber tones.

"Yes, sir. I certainly will," the general replied. He was stunned, staring at a large display that depicted the intercept point as a cartoon star-like explosion. A green arc representing the second Arrow's flight path had steepened as the unused missile dived toward Earth, tracked by the same radar systems in England, southern Europe, and Israel. Its warhead would automatically de-arm at some point, but where the missile would impact was uncontrollable. *At least it won't explode,* the general concluded. Squeezing the phone between a cheek and shoulder, he covered his eyes with his right hand, put his left hand on his yarmulke to make sure it had not slipped, and whispered to himself a thankful but joyous *sh'ma.* He knew that the Prime Minister was doing the same.

"Document the debris and that second Arrow's trajectory," the general ordered, one hand covering the phone's mouthpiece. He waited until two translucent red areas appeared on-screen, one for each of the projected impact zones. *Damn!* he swore under his breath. Both zones were in Jordan.

"Mr. Prime Minister?" the general asked, not certain the Israeli official was still on the line.

"Yes?"

"Sir, be advised that debris from the intercept, as well as the second Arrow missile, are projected to fall in western Jordan. Until proven otherwise, debris from the Shahab must be considered hazardous."

"You mean radioactive, I presume," Baron sighed.

"Yessir."

A long silence preceded Baron's reply. "Thank you, General. I'll take it from here."

"Sir . . . one question, if you please?"

"Go ahead," Baron ordered, slightly irritated.

"Will Israel respond to this attack?" the general asked. Cheeky, perhaps, but he had to know.

"Absolutely. Speedily and soon," the prime minister said, without hesitation, then signed off.

WHITE HOUSE SITUATION ROOM

"Mr. President, I'm sorry, but the prime minister is unable to speak to you at this time. He's off site, with his advisors. And extremely busy."

President Pierce Boyer's eyebrows arched in disbelief, as he threw palms-up hands at the ceiling. Glaring at the Situation Room's tele-

conference speaker-microphone device embedded in a huge conference table, Boyer fought for control before responding evenly. "I understand. I certainly do. But it's imperative that I speak to Prime
Minister Baron *before* Israel answers this unspeakable act of aggression by Iran. Please convey that urgency to him. *Please!*"

Boyer leaned on the table and locked eyes with T. J. Hurlburt, his
SecDef, who was seated on the opposite side. Herb Stollach and
Paul Vandergrift stood behind Hurlburt. All four were in shirt-
sleeves and sleepy-eyed, having been rousted from bed.

"I'm terribly sorry, President Boyer. But that simply is impossible . . . at least for now. I'm sorry," a heavily accented voice said.
There was no sense of compromise in its stern tone.

Boyer's head drooped, resigned. "Of course. Please convey my personal concerns to the prime minister, and ask him to call me as
soon as possible." Not waiting for an answer, Boyer angrily punched
an OFF button on the teleconference transceiver, then slammed a
fist on the table.

"The son-of-a-bitch wouldn't take my call!" he stormed. "He
wouldn't talk to the *President* of the United States! Who the hell
does Baron think he is? *Who* sends his country billions of dollars
every year, and *who* . . . ?"

"Hold your damned horses, Pierce!" Hurlburt interrupted
sharply, abandoning any semblance of White House deference. The
retired Army four-star general was tired, grumpy, and devoid of
patience with White House political nonsense. He wasn't about to
indulge Boyer's self-righteous tantrum, uncharacteristic though it
might be. Normally, Hurlburt admired the president's wealth of
Virginia-bred self-control. In fact, he envied Boyer's slow-to-boil
demeanor, a virtue Hurlburt could never claim.

Boyer's look of shocked disbelief was priceless, Hurlburt thought.
"Okay, okay! Hold your damned horses, *Mr. President*," Hurlburt
growled, over-emphasizing the title, then rushed on. "Look, if a nuke
had just been fired at Washington, would you *really* feel obligated to
drop everything and talk to Moishe Baron, especially if you knew
he'd only try to talk you out of shooting back? Christ, man! Baron's
got the entire Knesset and the whole damned IDF general staff

screaming at him to retaliate, probably with nuclear weapons! Let's give the man time to work his own national defense issues . . . *sir!*"

"Oh come on, T.J.! Nobody stiffs America's president, especially Israel. They owe their very existence to us! Baron's behavior is simply intolerable!" Vandergrift huffed, glancing at Boyer for support.

Slamming the table, Boyer shouted, "Shut the . . . *Just . . . shut . . . up, Paul!* God, will you ever stop yapping at *precisely* the wrong time, saying *precisely* the wrong thing? You really get on my nerves, mister!" Aiming a finger at Vandergrift, Boyer was clearly pissed, fighting for control, Hurlburt thought, silently gloating. Paul Vandergrift should have been canned long ago. Maybe the president had finally had enough of his esteemed National Security Advisor.

Vandergrift's demeanor changed immediately, recoiling as if he'd been slapped. Jaw and fists clenched, the tall man spun on a heel, crossed his arms, and leaned against the wall, glaring at Boyer. He was sulking, but finally had sense enough to shut up. Hurlburt smiled, staring at his own folded palms to avoid also incurring Boyer's wrath.

The president straightened slowly, jammed both hands in his slacks' pockets and walked to the front of the room, back to the others, also ignoring the large display before him. Television news images rocked crazily, indicative of a cameraman either trying to shoot on the run, or being jostled. Indeed, faces swirled before the camera, some accompanied by a fist thrust at the lens, others shouting noiselessly. Mercifully, the TV's audio had been muted. But the images alone were disturbing, Stollach thought. They were beamed from Jerusalem, where Jews and Arabs were, again, battling each other in bloody confrontations.

The entire Middle East is on the verge of disintegrating, Herb Stollach reflected glumly.

Hurlburt felt an incessant vibration over his left breast. Choosing to commit an act absolutely forbidden, during a Boyer-White House meeting, the SecDef reached into an inside coat pocket and retrieved his handheld communicator. Its screen displayed a "Flash" message from a familiar Strategic Command e-address. Good. Howard Aster was responding, and not a second too soon. Hurlburt scanned the text message, quickly grasping its gist and tone, before slipping the device back in his pocket.

Boyer turned and pointed at Hurlburt. "T.J., what's Baron going to do? What would *you* do in his position?"

"No question. He'll strike back, Mr. President," Hurlburt said quietly. Boyer moved closer, as if to hear better. "Hell, he has no choice! He can't let Ahmadinejad get away with damned near carrying out his long-standing threat to wipe Israel off the map. Baron *will* respond, and soon. He can't risk Iran lobbing another nuke at Tel Aviv. The next one just might get through. And . . . well, Baron doesn't have an overabundance of Arrow interceptors, does he?"

Boyer turned to Stollach. "Herb?"

Stollach flicked a hand at the massive screen dominating the room's main wall. "Take a look, sir. Israel's ready to blow wide open. Jews are beating up Arabs, and you can bet every Palestinian nut-job in Gaza and the West Bank is strapping on his explosive vest, ready to finish the job Iran *tried* to do. I'm with T.J. Baron will strike back, and he'll do it sooner rather than later."

Boyer pointedly ignored Vandergrift, turning and pacing slowly again. Over his shoulder, he asked, "What are STRATCOM's wargamers saying, T.J.?"

"Just got a message from General Aster, sir," Hurlburt answered. "His 'gamers assessed several scenarios and came up with the same answers every time: Israel will shoot back, almost assuredly with nuclear weapons. Probably within hours, not days. There's considerable disagreement about likely targets in Iran, though."

Boyer nodded, resigned to the obvious. "Then we'd better figure out how we posture ourselves for the fallout of Israel's retaliatory strike, gentlemen," the president said softly, his tone betraying a deep sadness. "The whole damned world's going to blame us for not stopping Moishe Baron, and that'll translate to yet another blowup in the Muslim world. I'm not sure there's a man or woman alive who can predict what will happen then, or how this mess will end."

Looking at Hurlburt, the president added, "We could be attacked anywhere in the world now. And probably will be. Keep our frontline units at THREATCON DELTA. In particular, make damned sure the Navy's primed and cocked for *anything* in the Persian

Gulf, especially in the Hormuz Strait. That Shahab may just be an opening shot of a larger Iranian campaign."

MISSILE LAUNCH SITE/ISRAEL

The massive, two-ton concrete disk pivoted slowly, revealing a deep, cylindrical cavern. Below ground level, a blunt-nosed missile had waited for years, ready to deliver its deadly cargo. Today it would answer Israel's call to duty. A black composite nose-shroud capping the Jericho 3 ballistic missile concealed a twenty-five-kiloton fission-type nuclear warhead. Alerted by electronic orders, the warhead's inertial guidance system automatically started refining its present position and preparing internal memory chips to load new target coordinates.

Miles away, two launch-control officers in an underground bunker worked their way through a detailed command-response, prelaunch check list, ensuring the Jericho 3 was ready to fire. Both were in their twenties, yet seasoned beyond their years. From birth, they had been immersed in a siege mentality that characterized their nation, acutely aware that the country was always at risk of being attacked and overrun by enemies surrounding Israel. Today, their nation *had* been attacked, spared only by God's grace and the technological miracle of an Arrow interceptor.

"Launch control, Jericho four-five. Prelaunch checklist complete; standing by," the ranking launch officer radioed. A thin, boom-type microphone extended from a cup-like headphone to the edge of his lips.

"Prepare to fire on my command," a deep voice replied. "Target coordinates are loading now." Through buried fiber-optic cables, a series of digital commands flashed to the remote missile silo, where they were fed to the Jericho's guidance and control computer.

Back in the launch-control bunker, a young captain glanced at his partner, a dark-haired, female lieutenant. "Are you prepared for this, Lieutenant?" the senior officer asked formally, one hand covering his microphone.

"Yes. Absolutely," she responded firmly. Her dark eyes were

wide, their pupils flared, a mixture of fear and anticipation familiar to combat veterans.

The captain recalled his own first combat engagement and smiled thinly. "Do not worry. You will do your duty," he assured.

"Stand by . . . ," a voice in their headsets demanded. "Jericho four-five, *fire one!*" The captain and lieutenant turned their keys simultaneously, closing a fail-safe circuit, sending an authorization-to-launch command to the missile. With his other hand, the captain lifted a plastic cover and thumbed a red switch forward.

The Jericho 3 instantly sprang to life, its rocket engines belching fire, smoke, and thundering noise. The missile momentarily strained against explosive tie-down bolts, then was freed. It initially ascended slowly, but accelerated quickly as it exited the launch silo. Standing on a compact tail of white-hot fury, the Jericho 3 streaked into the sky, trailing thick white smoke.

"May the Lord guide you home," the young lieutenant whispered, monitoring a missile-status display.

Over the next few minutes, the missile's solid-propellant stages burned fiercely, flamed out, and separated. Soon, the composite nose shroud departed, leaving a mean-looking conical warhead to arch over and dive toward Earth from the dark of space. The warhead vibrated slightly as the atmosphere thickened. Soon, supersonic shock waves stretched aft from its sharp nose tip. Tiny thrusters fired, adjusting the warhead's trajectory and guiding its cargo to the intended destination with four-decimal-point accuracy. One guidance thruster refused to respond, a casualty of age, but the others compensated for its failure.

"Tracking radar, report!" the Israel Defense Force launch commander demanded.

"On track; in the basket," an operator barked. His computer displayed an imaginary, translucent-red funnel with its tip on the target, fanning wider as it extended upward, into space. A thin green arc representing the Jericho 3's ballistic path was now diving into that funnel, dubbed by missile professionals as "the basket."

The Israeli commander nodded, his gaze locked onto a shock-mounted monitor affixed to the concrete bunker's wall.

At 0837 local time, the Jericho 3's hardened, penetrating warhead dove into the sands of Iran's desert, plowing through forty feet of compressed soil and three feet of steel-reinforced concrete. On-board accelerometers sensed when the warhead entered a cavity, triggering a massive detonation inside a large cavern filled with equipment and people. Several hundred Iranian technicians, engineers, and scientists literally evaporated in the initial nuclear blast. A thousand centrifuges busily refining smuggled uranium into weapons-grade fuel were melted or crushed into twisted lumps of metal. The detonation's tremendous shock wave and resultant fireball fractured concrete structures throughout the facility, smashing and incinerating everything they encountered.

At the surface, desert sand and rock heaved, spewing dense fountains of soil, chunks of ugly concrete, and steel hundreds of feet into the air. An escaped shock wave raced across the barren landscape in all directions, slamming into workers' homes, small shops, and schools a kilometer away, instantly reducing them to rubble. Close behind, a hellish firestorm swept over the shock wave's destruction, sucking oxygen from crushed lungs and igniting wood-frame buildings, clothing, shopping bags—anything combustible.

In seconds, Iran's sole uranium-processing facility was gone. Hundreds of workers died, buried in the underground plant by tons of sand, rock, steel, and concrete that plunged back into the massive pit created by the initial explosion. In the nearby city of Natanz, workers' families joined their breadwinners on the other side of death's veil. No single person had seen or heard the incoming devastation. Each was killed in seconds, death snatching life from their bodies without warning.

Dust and smoke soon swirled in the ghostly quiet of Natanz. No voices, no moans of pain, no barking dogs. The thunk of a cinder block or roof tile falling from a shattered structure sounded now and then, its noise muffled by grit-filled air. No human was alive to hear them, or to witness the angry, huge cloud of dark smoke and sand climbing skyward, appearing to curl inward on itself, yet expanding without bound.

Far above, in geosynchronous orbit, nuclear detectors on America's Space-Based Infrared System and aging Defense Support Program missile-warning satellites logged the flash of visible light and the flood of invisible gamma rays. Other "nudet" instruments on GPS satellites in mid-Earth orbit also recorded a nuclear event at Natanz. Alerts from those spacecraft, backed by ground-sensor data, immediately appeared on display screens at Air Force Technical Application Center headquarters in Florida. Minutes later, aural alerts and digital messages were fired from AFTAC to STRATCOM's command post, the Pentagon's National Military Command Center, White House, NORAD, and other key sites across the globe: *Nuclear detonation at Natanz, Iran. Source: Israeli Jericho 3 missile/warhead. Details to follow. . . .*

PRIME MINISTER'S OFFICE/TEL AVIV, ISRAEL

"And you have confirmation?" Moishe Baron asked sternly, speaking into a mobile handheld communicator.

"Absolute confirmation, sir," an Israeli general officer assured. "Our Ofeq-11 imaging satellite was in position to monitor the impact and subsequent destruction at Natanz. The images are degraded somewhat by radiation damage to the spacecraft's electronics, but they're of sufficient quality to provide airtight confirmation. After digital processing, the pertinent images will be forwarded to your office."

Baron wanted to ask about the nearby city of Natanz, where the nuclear facility's personnel and their families had lived, but he already knew the answer. He'd seen the prestrike simulations, and knew nothing within miles could have survived that twenty-five-kiloton blast. The damned Iranians had built their underground uranium-processing and weapon-assembly facility much too close to that city. Clearly, Ahmadinejad and Iran's fanatic imams had little regard for the safety of women and children, a fact that initially had given Baron pause. But he, as prime minister of Israel, had still given an order to destroy the nuclear plant, knowing hundreds, maybe thousands, of innocents would also die in the nuclear holocaust.

Baron thanked the IDF general and asked him to convey the appreciation of Israel's people to the missile-warriors he commanded. Once again, the nation's powerful military force had protected its fellow citizens by destroying a nuclear threat in the region.

Snapping the communicator's cover closed, Baron stared briefly at a picture of his wife and children. For now, they and all other citizens of Tel Aviv were safe. But for how long? *Will Israel ever live in peace?* he asked himself, indulging in a flash of near-despair.

"You made the right call, Moishe. There was no other way." Wearily, Baron looked at the man seated on the other side of a broad desk. Compact, hard-edged, and absolutely fearless, Ezir Lichtenstein fit the stereotype of a career Mossad agent. Better known by his nickname, "Ossi," he now commanded the storied clandestine intelligence service. Ossi had earned his top post the hard way, ferreting out critical information from countries across the globe, and killing those who had threatened his beloved homeland. Under his leadership, Mossad had grown in size and effectiveness during recent years, and had cemented relationships with intel agents from other countries. In short, Mossad had become more intimidating than ever.

"I know," Baron nodded, a wave of exhaustion rolling through his being. "But it doesn't diminish the fact that . . . *Damn!* So *many* senseless casualties . . ." Baron's voice cracked, forcing him to turn away.

"Not so many, I think," Lichtenstein offered, displaying no emotion, let alone empathy for Baron's distress. "How many thousands of Jews would have died, had Iran's Shahab hit Tel Aviv? You must think of *our* people, not the heathen who were building *more* 'Islamic Bombs' to kill Jews!"

Baron sighed and agreed, glancing again at the photo on his desk. He turned to face the Mossad chief and cleared his throat. "It is what it is, Ossi. Now. Are you prepared to immediately launch the next phase?"

"Of course, Moishe. Most definitely. And immediately!" Ossi was nothing, if not self-assured. That fearless confidence had paid off for him time and again. But, across the decades the two men had served together, that smug *uber*confidence had always irked Baron. In the prime minister's experience, nothing was certain. As the crass Americans put it, shit *did* happen. But he also knew such bravado and can-do attitudes were legion among Mossad's shadow warriors. Maybe it was necessary for them to do the work most could not or would not do.

"Then proceed. As soon as you confirm the targets are together," Baron ordered.

Lichtenstein tipped his head in acknowledgment and smiled.

That almost-smirk was very similar to the squint-eyed smile exhibited by Iran's irrational president, Baron mused.

"We have the means to determine that with great precision, Mr. Prime Minister. Consider it done." The Mossad chief stood and offered a hand. Baron gripped the outstretched palm for a long few seconds. Neither man spoke. The gravity of their undertaking was well understood by both.

A HOTEL ROOM/TEHRAN, IRAN

Mo Barak stood well back from the dingy, streaked-glass window, so he would not be seen from the street five stories below. With a pair of small binoculars, he could easily monitor activities a block or so away, where a helicopter had just landed in a large courtyard of Iran's presidential palace. The sharp viewing angle was less than ideal, but sufficient to identify a tall, thin figure that exited the aircraft's left-front door, then stooped to remain clear of rotating rotor blades as he trotted toward an arched doorway. The deep-cover Mossad agent smiled as his binoculars tracked Hassan Rafjani.

Perfect. They'll be together in a few minutes, Barak was willing to bet. The encrypted message he'd received from Mossad headquarters a few hours earlier made it clear that he, Barak, should *not* act, unless both Iran's President Ahmadinejad and his powerful security agent, Hassan Rafjani, were together. Given recent events, Barak suspected the two men would soon be in heated conversation . . . and in the same room.

The agent pulled an electrician's tool kit from a doorless closet and dumped its contents on the frayed, faded carpet of his second-rate hotel room. With practiced speed, he removed a false bottom from the scarred leather-and-aluminum case and carefully retrieved two small electronic devices from their black-foam packing. He twisted the larger box, which separated a six-inch-square, half-inch-thick plate from a larger section. The two were still attached by a coil of insulated wire.

Barak set the plate on a meter-square table near the window and aimed its segmented front surface at the building Dagger Rafjani had just entered. The Mossad agent stooped to sight across the top of the plate, adjusting its position slightly, before connecting the other section to a thin notebook computer. Seconds later, a green light-emitting diode told him the two-piece device was operating properly. He entered an alphanumeric code via the computer's keyboard, and was rewarded by a hazy gray-scale image on the screen.

Satisfied, Barak patiently powered up the remaining device, unfolded a triangular-shaped antenna, and placed the box on a paint-chipped windowsill. He reviewed a memorized checklist, adjusted several parameters via the computer, then studied the image on its high-resolution screen. Barak rubbed his hands together, anticipation running high now. Thanks to this wall-penetrating, ultrawideband-radar system, he was monitoring activities inside the office of Iran's president, Mahmoud Ahmadinejad.

He'd spent many hours watching the interior of that room, carefully tweaking the radar's "range-gate" parameters to exclude objects on either the near or far sides of that office. The magic of ultrawideband radar's super-short, extremely covert pulses had given Barak and other Mossad agents a ringside seat to the interior of many such rooms in recent years. As a result, uncertainties about the presence—or not—of targets had been reduced substantially, greatly increasing the secret agency's mission success rates.

He waited until the display showed a tall figure Barak recognized as Rafjani joining a second, smaller figure in the office. Seconds later, a flock of heavyset figures seemed to storm through the president's office doorway. Barak didn't know who the others were, but he suspected, based on their bulk, that these were the principle imams who constituted Iran's theocracy.

Most of the nation's esteemed senior imams loved to eat, and their girths were testimony to frequent indulgence. Barak perused the computer screen closely, unable to believe his incredibly good fortune. With just a bit of luck, maybe the old goat himself, Supreme Leader Ali Khamenei, was with that group. There was no way to be sure, but, clearly, something big was taking place in the president's office, given the number of figures now clustered near Ahmadinejad and Rafjani.

PRESIDENT'S OFFICE/TEHRAN, IRAN

"You are a fool, Rafjani!" Khamenei bellowed, thrusting a bony finger at the target of his fury. "Your misguided attack on Israel was yet another miserable failure! And it has brought great shame on Persia and its holy leaders—*those who stand before you now*! How dare you launch a Shahab carrying Persia's last nuclear device without consulting this council and your president!"

The old Ayatollah continued to rant and shout, but Rafjani tuned him out. *Idiots! All idiots! I will kill all of you fainthearted "holy women," beginning with you, old man*, he decided coldly. He glanced at Mahmoud Ahmadinejad, but the president's face betrayed nothing. Iran's president smiled the cobra's smile, politely nodding now and then as he pretended to listen to the Ayatollah. Finally, the old man waved a withered hand and turned away in disgust, his tirade ended.

But Dagger's tongue-lashing was not yet complete. The Secret Intelligence Service chief, a rotund man with a thick, gray beard, continued. "Israel has *not* been destroyed, as you promised, Rafjani! Only through *your* gross incompetency could the Jews' puny interceptors shoot down our holy, invincible Shahab! Once again, Islam's might appears to be inferior to Zionist technology, at least in the eyes of Muhammad's people. On your head fall the hopes and dreams of all Muslims! You have disgraced yourself and your people!"

A block away, Mo Barak was not privy to the dressing-down Rafjani was receiving, because he only saw blurred images on his computer screen. But what he could discern was enough to decide the time to act was *now*. He reached across the small table, trying to steady a quivering hand, and pressed a tiny rubber-covered switch embedded in the top surface of the compact transmitter box. At the speed of light, a burst of coded ultrawideband pulses raced across the few hundred feet between Barak's hotel room and the Iranian president's office.

Inside that office, secreted beneath two standard light-switch plates near the room's two doors, tiny receivers awoke and captured that stream of UWB ones and zeros. Barak had installed the nanoscale receivers months before, posing as a contractor technician performing routine electrical maintenance. The two receivers, in turn, commanded compact, powerful transmitters to fire bursts of intense electromagnetic energy into the room. The two devices' fan-shaped patterns blanketed the room's interior, hitting every man there.

Rafjani had just risen to counter the blather spouted by the old imam and SIS chief, when an intense, stabbing pain exploded in his chest. He gasped, unable to breathe, conscious of the most intense, searing pain he'd ever experienced. He dropped to his knees, both hands clawing at his chest, eyes bugged wide. He was unaware that President Ahmadinejad, the Supreme Leader Khamenei, the SIS chief, and every mullah in the office were suffering the same distress. All were gasping and clutching their chests, all alarmed at the excruciating pain. Ahmadinejad managed to take two steps before pitching facedown on the thick carpet, his mouth agape in a silent scream.

Rafjani knew he was dying. Somewhere, deep in a corner of his mind, part of him raged with white-hot anger. *No! This is not my destiny . . . !* The thought was fleeting and trivial, but also his last. Electromagnetic waves fired by the two transmitters hidden beneath wall-switch plates had stimulated abnormalities in his heart's electrical "spiral waves," inducing arrhythmia and fibrillations throughout the fist-sized, life-critical muscle. These triggered fatal disruptions of the heart's electrical controls and, ultimately, total heart failure. Dagger Rafjani fell forward, his face smashing into the floor, disbelieving eyes wide. Around him lay the room's other occupants. All were dead.

U.S. SPECIAL INTELLIGENCE REPORT (JUNE 2011)

IRAN: *By all indications, Iran's nuclear weapons program has been successfully interdicted. Parchin and Natanz, the nation's primary uranium-enrichment and weapons-development facilities, have been destroyed. Iran retains a sizable Shahab missile fleet, but probably no functional nuclear weapons.*

The unexpected deaths of Mahmoud Ahmadinejad, president of the Islamic Republic of Iran, and Supreme Leader Ali Khamenei have precipitated a rapid demise of radical Islam in Iran. Shock and uncertainty were the initial reactions of Iran's citizenry, which prompted an immediate clampdown on liberal and moderate elements by Republican Guards and their ultra-secretive Quds cadre. However, a groundswell of spontaneous protests among millions of common Iranians, who had grown weary of the aggressive international antics of President Ahmadinejad, the deteriorating national economy, and an increasingly repressive theocracy, quickly altered Iran's political landscape. Consequently, the Republican Guards have retreated to their garrisons and, for now, are taking a hands-off approach to national politics. However, the Guards and Quds remain major strategic concerns.

Moderate Iranian leaders are gaining favor with the populace, opening the possibility that a more-democratic regime may be in the offing. Career bureaucrats are keeping Iran running fairly well, despite the vacuum of political leadership. CIA analysts estimate that, within a year, Iran may no longer need to import gasoline. It is premature to predict whether these factors will improve stability or not.

SAUDI ARABIA: *Publicly, Saudi leadership has criticized Israel for its military response to Iran's nuclear* *attack, citing "excessive force" and "unacceptably high" casualties in Natanz. However, Saudi back-channel communications with the U.S. and Europe present a much different picture. Saudi royals are expressing relief that Iran's radical leaders are no longer in power. Arab elements are working closely with Iranian moderates to build a government more palatable to Iran's neighbors, the Saudis claim. A particularly unsettling issue, however, is this: Saudi secret police and military special forces teams are systematically eliminating radical Islamic cells throughout the Saudi Kingdom and Iran. Intelligence sources report that similar teams also are decimating Hezbollah and Hamas forces in Gaza, Syria, and Lebanon.*

SYRIA: *Radical-Islam elements within Syria are rapidly dissolving, following the loss of Iranian sponsorship. Top-level government emissaries have approached senior Israeli officials, communicating that Syria would be open to negotiating a comprehensive peace treaty with Israel.*

ISRAEL: *The Jewish state has reverted to its traditional bunker mentality, exhibiting considerable nationwide concern and worry. Citizens' angst and fear of another Iranian nuclear attack was mitigated somewhat by detailed news stories and photos of the destroyed Natanz facility. The subsequent shocking deaths of Iran's Ahmadinejad and Khamenei, which closely followed the Natanz strike, seem to have accentuated, rather than lessened, Israeli fears of "what might come next," with respect to Iran. The nation's defense forces remain on high alert, signaling continued government pessimism that the final chapter with Iran has yet to be written. Tel Aviv still considers Iran a potent threat.*

NOTE: *Through independent investigations, CIA and foreign intelligence agencies are finding considerable evi-*

dence to support what was simply rumor a few days ago: that six-to-eight of Iran's top leaders, including Ahmadinejad and Khamenei, died of massive heart attacks. All were in the president's office, and all apparently died at about the same time. A deep-cover CIA asset in Iran reports that Hassan "Dagger" Rafjani, a security agent close to the president, was one of those who suffered a fatal heart attack. Rafjani has been linked to the deaths of numerous CIA assets; was Iran's financial tie with a Colombian drug cartel responsible for disabling numerous U.S. satellites and killing five astronauts on the International Space Station last year; and was responsible for the brutal, televised execution of a Canadian woman, Charlotte Adkins, who was in Tehran as an unofficial emissary for the U.S.

At this time, the U.S. intelligence community has no credible explanation for up to eight senior Iranian leaders dying of massive heart attacks. The fact that they all expired together suggests the use of an unknown, highly lethal weapon. The National Intelligence Director certifies that there was no U.S. involvement in this incident. Investigations are continuing.

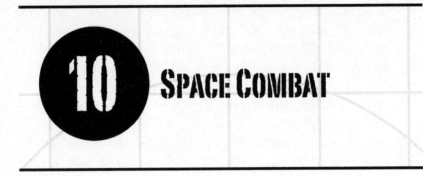

10 SPACE COMBAT

WHITE HOUSE/OVAL OFFICE

Secretary of State Austin "Mac" McAuley carefully returned a china coffee cup to its saucer, before concluding, "That's all we know about the situation in Iran, Mr. President. Things appear to be settling down, but a definite political power vacuum exists in Tehran. And the Republican Guards are still a wild card."

"For now, though, the Guards and their Quds cadre are lying low, correct?" President Pierce Boyer asked. He sat comfortably on an elegant, gold-threaded divan, legs crossed, facing McAuley and Paul Vandergrift, his national security advisor. The two men were in wing-backed armchairs, facing the president across a square, glass-topped coffee table.

"That's right. They've basically holed up in their barracks areas. Far as we know, they're staying out of the political brouhaha, at least for now."

"How can we be sure of that, Mac? Where's this information coming from?" Vandergrift asked, not attempting to conceal acute skepticism.

"Solid intel from State Department assets, backed by National Reconnaissance Office overhead imagery—admittedly degraded imagery—and from a deep-cover CIA agent in-country," McAuley replied.

"You're telling me *State* has assets in Iran?" Vandergrift asked, a mite too sharply.

"No, not in Iran. But State *does* have reliable sources in the region, I assure you." McAuley wasn't about to divulge too much concerning State's resources in the Middle East. He trusted Paul Vandergrift about as far as he could throw the pompous ass.

"Paul, drop it," Boyer barked. "State's diplomatic corps has always had its own regional networks. They serve as a valuable cross-check of national-security intel sources. And, so far, Mac, the intel community's reports on Iran are consistent with State's, as I read them.

"Another concern," Boyer continued, throwing an arm across the divan's arched backrest. "It sounds like our friend Chavez is ratcheting up his verbal attacks against us. Based on this morning's intel summary, it appears he's trying to blame the U.S. for Iran's turmoil. What's that all about?"

McAuley nodded, pulled a single sheet from a leather-bound folder and scanned it quickly. "Hugo's being a spoiled brat again, but our analysts believe he's not really tweaking *our* nose. This latest bluster is strictly aimed at Muslims, not the U.S. With Iran out of the picture, Chavez sees an opportunity to become the world's default leader of a few billion poor and oppressed Muslims. In his warped view, he's assuming the mantle of spokesman and caretaker for the world's downtrodden. He intends to pull all of Islam into his tent."

"To what end, for God's sake?" Vandergrift exclaimed.

"We're not absolutely sure, but our analysts say this latest campaign is related to a recent oil deal involving Iran, Russia, and Venezuela. The three nations secretly established a new cartel they call the People's Oil Consortium, to counter OPEC and drive up oil prices. As you know, POC was formed to disrupt U.S. and European economies through artificial oil 'shortages'—when they decided the time was right."

"That I can buy," Vandergrift said, nodding. "National Security Agency intercepts about this POC primarily involved that Iranian hardliner, Rafjani, a high-level Russian, and Chavez's top people.

Rafjani was one of the recent Iranian-leadership casualties, by the way. He appeared to be the architect of POC. Chavez must . . ."

"What the *hell* is that all about, Paul?" Boyer exploded. "What NSA intercepts? What new consortium? Why is this the first time *I'm* hearing about POC?"

McAuley started to speak, but was cut off by Vandergrift, who flipped a hand dismissively. "It wasn't important, Pierce. I didn't mention it because, frankly, I've had serious doubts about the veracity of NSA's reports of late, and I saw no reason to bother you with questionable information. Until that Iranian leadership decapitation . . ."

"Good God, Paul! How can the President of the United States make critical national-security decisions when he knows *nothing* about a new consortium that may control a goodly percentage of the world's oil? Especially one capable of upsetting the balance of power throughout the entire Third World?" Boyer stood and pointed at McAuley, ordering, "Mac, by tomorrow morning, I want a full assessment of this POC consortium and what it means to the U.S., particularly in light of the latest developments in Iran."

"Sure. I'll have someone here by nine tomorrow, sir," McAuley agreed, suppressing a smirk. Maybe Vandergrift's lease on a White House office was expiring, after all. While Boyer paced, obviously still boiling, McAuley quickly concluded State's assessment of Hugo Chavez's latest moves, then excused himself. As the Secretary of State slipped out the Oval Office door, it was clear President Pierce Boyer was verbally ripping a good-sized chunk of hide from Paul Vandergrift's posterior.

PENTAGON/SECRETARY OF DEFENSE'S OFFICE

"To summarize, we believe Iran is far less of a threat now, due to Israel's nuclear retaliation and the decapitation of Iran's leadership. We still aren't sure who to thank for the latter," SecDef T. J. Hurlburt quipped, facing a videoconference camera. His image was being beamed to U.S. combatant commanders scattered across the globe.

"President Boyer's not too damned happy with Israel, but, overall, the White House is mighty relieved that Iran's nuclear capabilities have been neutralized. Consequently, he's ordering all combat commands back to THREATCON BRAVO. That doesn't apply to ongoing operations in Central Command's area, of course."

The SecDef leaned toward the camera and folded his thick hands. "Look, we still have a lot of touchy situations across the world, so don't *totally* stand down. You all know how we're hurtin' in the space arena. Every bad actor on Earth is acutely aware that America's space-based capabilities were severely compromised over the last year. Consequently, a few of those renegades are feelin' their oats, and they smell an opportunity to cause mischief. I assure you, folks, the United States is *not* as blind, deaf, and mute, in a space sense, as those bad guys think we are. But perception is reality in North Korea, China, Africa, South America, et cetera, so we have to be ready for anything, anywhere."

Hurlburt raised both palms, opening the line to dialogue. "Comments? Concerns?"

"Excuse me, sir," interjected the head of Pacific Command, a U.S. Navy admiral. "We still have bombers in Guam armed and on high alert, ready to launch. Given the unpredictability of North Korea and the force movements we see in China, I'll be keeping at least half of those birds cocked and ready. Overhead imagery and signal intelligence data from both Korea and China are marginal at best these days.

"Consequently, we're maintaining a conservative posture by keeping a quick-response force on alert, ready to counter any threats from those two. That may require a plus-up of Air Force crews in Guam, sir, to ensure we can sustain a high readiness level over the long term."

Hurlburt and the other videoconference participants discussed the Pacific commander's comments, trying to balance his personnel needs with those of other regions. "Basically, we have more requirements than bodies," Hurlburt finally concluded. "Folks, we've lived with this too-few-doing-too-much challenge since the Cold War ended, and I don't see it getting better anytime soon, given the nation's ongoing financial crisis. That's why it's absolutely essential we take care of our troops. So, make the tough calls on readiness, but make damned sure you give your people some downtime."

He pointed at one of the screens before him, adding, "Howard, that applies to your wargamers, as well. Keep Stanton Lee's China-related side-game going, 'cause it's time-critical. But I want most of the other 'gamers to take a break. That means you and your battle staff, too, General," Hurlburt growled, half-grinning.

STRATCOM HEADQUARTERS/COMMANDER'S OFFICE

General Howard Aster studied a PowerPoint summary slide, then asked a few questions of a shaved-bald Army lieutenant colonel, who had presented the National Security Space Institute team's briefing. Aster listened closely to the answers, nodded and thanked the officer, then turned to Major General Dawn Erikson. She sat to Aster's right, wearing a dark skirt, pale-blue blouse, and Air Force dress uniform jacket, complete with rows of ribbons topped by a space badge. Dual silver stars sparkled on each shoulder epaulet.

"Viking, I'm impressed. Outstanding concepts! You space-warriors have done a great job," he said, a sweep of his arm including the lieutenant colonel briefer and two other officers seated at a conference table. "If Fourteenth Air Force's conjunctive analyses determine Chavez's VenezSat *might* constitute a threat, we now have viable options for dealing with it."

"The credit goes to these guys, sir," Erikson said, nodding to the other officers. "Our institute's staff has some of the smartest space cadets in the national security community. These guys are 'what-iffing' all the time, either in the classroom or after-hours bull sessions."

"Good on 'em! It's refreshing to see innovative thinking about this cutting-edge warfighting arena. Combatant commands, like STRATCOM—and ol' fighter pilot commanders like me—are out of our element when it comes to the space battle arena. America's in dire need of a space 'think tank,' and NSSI is damn sure on-track to fill the future-gazing role," Aster enthused.

"You folks are today's Billy Mitchells of milspace," he proclaimed, referring to the World War I–era Army officer now regarded as a visionary air-power pioneer. The STRATCOM four-star glanced to his right and grinned slyly at Erikson. "Maybe I should keep your team here, just to keep my space-butt out of trouble!"

She laughed, a soft tinkle that reminded Aster of fine crystal. "Wish we could, sir," she said evenly. "I'm afraid we'll have to leave first thing tomorrow, though. NSSI is hardly overstaffed, and we have a new 'Space 300' class arriving next week. These guys are my primary instructors for that course, and they need a day or two to update their presentations. That nuke-weapon exchange between Iran and Israel dictates a major reworking of NSSI's module on missile defense."

"Understand," Aster conceded, but his jaw tightened. "The Pentagon says things have calmed down to the point that THREATCON levels are being lowered. Frankly, I have serious doubts about that assessment, given our perilous space infrastructure. But Space Command's working hard to restore the most-critical capabilities. Anyway, I may require some quick-response NSSI assistance, okay?"

"Of course, sir," Erikson assured. "We'll be glad to come back."

Effecting a conspiratorial tone, he raised an eyebrow and flicked a thumb toward Erikson's U.S. Army aide. "Surely you could spare Hans for a few days. I could use his no-bull insights, as we work our way through this VenezSat issue."

Major Hans Richter glanced to Erikson before shaking his head. "Sir, I'm also part of NSSI's one man-deep Space 300 instructor cadre. Besides, if I don't get back to Colorado Springs, my bride will pitch a bunch of Army uniforms, rucksacks, and GI boxer shorts out the door, then file for divorce," Richter quipped.

"She hasn't seen Hans since the LAX incident," Erikson explained, "and is obviously worried about her Ranger."

"I hear you, Hans. Even Rangers' kitchen passes expire at some point, I s'pose." Aster chuckled. He stood and circled the table, shaking hands with each NSSI officer. "Gentlemen, a big thanks to each of you. You've provided high-value insights for our wargamers, and I damn sure appreciate your help." He touched Erikson's elbow and added, "Viking, could you stick around a few minutes? I need your input on another matter." The other three officers took the hint and filed from the office. Richter, still on crutches, appeared to have mastered them, Aster noted.

Once they were alone, Erikson cocked her head and smiled up at Aster. "So you need *my* input on another matter, huh?" Her eyes had that impish sparkle, which had caught Aster's attention during the 2003 war in Iraq.

"Well, I . . . Aw, hell! No, I don't *need* your 'input,' General Viking," Aster scoffed. "We just haven't had a minute to ourselves since you got here! You really have to bail out tomorrow?" His extra-attentive gaze betrayed a twinge of pain, Erikson thought. That surprised her.

"I'm afraid so, sir. I think we've done about all we can here, at least for now. The NSSI is still evolving, and part of my job is to serve as ambassador for the institute, helping Washington types and our coalition partners learn about our mission. Several congressional staffers are inbound for critical pre-budget briefings this week. I need to be on-site to help shepherd them. And . . . I'm a reservist, you know. I have non-Air Force responsibilities, too," she added cryptically.

"Like taking care of a family?"

The blunt, personal question caught her off guard. "Well . . . no. Not a family. A husband, yes. He's in Europe on business, and he was pretty shook up by the LAX thing."

A flicker of disappointment flashed across Aster's face, she thought. He nodded, letting a silence hang between them for a long beat.

"I knew you'd married, but . . ."

"No kids. Not at our age," she interjected quickly. Her expression softened to one of furrow-browed compassion, as she placed a hand on Aster's forearm. "Sir, I heard about Angie. I'm so sorry . . ."

Aster's smile vanished and his jaws tightened. He stared deeply into Erikson's eyes, then slowly nodded. "Thanks, Dawn. It's been two years since she died. Ovarian cancer. Found it too late."

Erikson's expression-rich eyes radiated sincere empathy—and an unsettling, knowing sadness, Aster thought. Those icy-blues had pierced him from the first time he'd met her, back in the Qatar desert. Today, they penetrated to his inner being, triggering an uncomfortable, way-too-exposed sensation. Other than the odd flash of anger, Aster rarely allowed his emotions to show. He was a product of the old-school fighter-pilot mentality that said real men flew and fought and raised hell. They damned sure didn't get weepy over personal matters, regardless of how much hurt lived deep inside.

But now, this pretty woman's knowing eyes sliced through his time-hardened, protective, emotional shield and freely roamed his soul, grasping the pain; sense of loss; and empty, aching loneliness that had become Aster's nightly companions. He turned away from those lovely, sympathetic, downright scary orbs and stared out the window a long moment.

"I'm sorry, Howard . . . uhh, *sir*. I didn't mean to . . ." Erikson stammered.

Aster laughed softly, dissipating the uncomfortable atmosphere. "Forget it, Dawn. I appreciate the kind thoughts." He faced Erikson and lightly gripped her upper arm. "Listen, about that 'other matter.' If you're intent on headin' back to the Rockies tomorrow, I'd be honored if you'd let me take you for a flight this evening—and to dinner."

She hesitated, again unbalanced by Aster's abrupt change of subject. It was his way, she knew, but the rapid shifts always caught her off guard. "Just to say 'thanks' for all your help here," he added quickly, offering that crooked, alluring half-grin. "Besides, the SecDef practically ordered the STRATCOM commander to lighten up a little. How 'bout it?"

"Well, to paraphrase a great movie, you had me at 'I'd be honored.' Yes! I'd love to go flying with you—especially if you're offering an F-16! And I'd love to have dinner with you, General," she added, with mock formality.

"No F-16, I'm afraid. Just my Cirrus bug-smasher, but it's a great perch for watching a sunset."

His desk phone buzzed angrily. Aster grimaced at the blinking light. Annie must have some heavy hitter on the line, or she would never have disturbed two general officers "in conference." Erikson pointed at the light and turned to leave the office. "What time, sir?"

"Eighteen-hundred. I'll pick you up at your quarters," he replied, reaching for the phone. "Casual dress! We're talking airport fly-in fare, not five-star dining. And let's give the aide and exec a night off, okay?" Nearing the office door, she turned, waved an affirmative and flashed a bright smile.

Aster waited an extended second before lifting the receiver, eyeballing well-sculpted calves. *Dawn, you are one hell of a woman!*

AERO CLUB FLIGHT LINE RAMP/OFFUTT AFB, NEBRASKA

"What a cute little plane!" Dawn Erikson exclaimed, as she and Howard Aster approached a Cirrus SR22-G3 four-place light aircraft. As a favor to the STRATCOM commander, the Offutt AFB aero club's manager had pulled the blue-and-white, single-engine, composite plane from the club's aging hangar and tied it down on the expansive concrete ramp. The aircraft now faced taxiway "Mike," which led to Offutt AFB's main runway.

"Hummph!" Aster snorted. "I have to confess, Viking. 'Cute' is a term I've never associated with a three-hundred-and-ten horsepower, turbo-charged aircraft that'll zing along at two-hundred-and-nineteen knots true airspeed!"

She wrinkled her nose at him and tossed her head. Dressed in snug jeans, a loose-fitting blue pullover top and low-heeled, slip-on shoes, she could have passed for a college coed. "Well, it's not an F-16, but I guess it'll do. This is *your* airplane?"

"Yep," he responded proudly, unlatching the left door and tossing a kneeboard onto a cream-colored leather seat. "The Air Force doesn't let old four-stars fly F-16s, at least not without some wet-nosed instructor pilot to babysit him. So, after Angie died, I bought this baby to help get my head squared away. She never liked little airplanes, but flying's been such a big part of my life. . . . Nothing like cruising over the prairie at twenty-five thousand feet to get your mind off things you can't do a damned thing about."

"I see," she said, not sure how to answer.

"Come on. I'll give you a rundown on this 'cute' little bird, while I preflight it." As Aster checked the oil level, ran a hand along each propeller blade, and scrutinized the leading edge of each wing, he chattered away, enumerating the finer points of his SR22-G3.

Clearly, the STRATCOM commander was transforming, shedding his four-star, on-duty persona and becoming an ordinary man passionate about flying. Erikson nodded politely and asked a few questions, but wasn't really paying attention.

Instead, she studied the man's face and smooth movements, focusing on tanned, bare arms, slim hips, how the evening breeze rearranged his hair. He'd been under tremendous pressure the past few weeks, she knew, and there would be more long, tough days ahead. It was nice to see him relax. The fading early summer light softened crow's-feet creases that she'd noticed around his eyes a few days ago, when he'd met her at the Offutt AFB transient-aircraft ramp. His hair was completely silver-white, not the salt-and-pepper mix she'd remembered from their time at the Air Operations Center in Qatar. He was eight years older, but, if anything, even more attractive than she remembered.

Completing his quick, but thorough, walkaround inspection, Aster tugged up a leg of stonewashed blue jeans, stooped, extended an arm under the right wing and unhooked the last of three tiedown ropes. Erikson took note of firm, well-defined, sinewy shoulder muscles bulging beneath his stretched-tight yellow golf shirt, a sight that triggered an unwelcome warmth and yearning. As he stood, she turned away, trying to swallow a catch in her throat.

Oblivious, Aster placed a hand in the small of Erikson's back, gently guiding her to a step near the aft edge of the SR22's right wing. "Hop up on the wing there, and I'll help you get strapped in. Stay on that black strip next to the fuselage. Can't have you punching a hole in the skin!" he laughed easily.

With his assistance, she climbed onto the wing, opened the right cabin door, and slid into a leather front seat. The material was surprisingly soft and luxurious. Warmed by the day's rapidly departing sunshine, the cockpit exuded a pleasant new-car aroma.

Aster knelt on the wing next to her seat, the right door held open by a knee. "These two go over your shoulders," Aster explained, arranging the shoulder straps. "They have airbags built into them, just in case. . . ."

Erikson fumbled with the left lap belt, while Aster threaded clips on the ends of both shoulder straps onto the right belt's metal, bayonet-like clip. He took the other strap's buckle from her, snapped the mechanisms together, and gently tugged the seat straps snugly across the woman's hips.

With his body so close to hers, Erikson detected a hint of familiar aftershave, a masculine scent that triggered a flood of memories. In a flash, it transported her back to another time and place. She and Aster were standing close, poring over reconnaissance images, arguing whether a target had been destroyed or not. His arm pressed against hers, his scent making it hard to concentrate on Iraqi targets, her heart pounding . . .

The memory and the moment were overpowering. She lay a hand against Aster's cheek, turned his face to hers and kissed him, holding his lips with hers. He didn't resist. To the contrary, he returned the favor for several luscious, long seconds. She gently pulled back, but held his face close, her eyes only inches from his.

Aster said nothing, holding her intent gaze, wishing with all his heart that his eyes could focus on those beautiful ice-blues. Age had degraded his near-vision to the point that, without "cheater" reading glasses, he couldn't focus on a damned thing up close. But it didn't matter. He could sense the softness, the deep feeling. Neither spoke for another extended minute, a frozen moment of stirring, tumultuous emotion.

"I shouldn't have done that," Erikson finally said, her voice shaking, barely under control.

"Am I complaining?"

She smiled, eyes dancing across his. He was acutely aware that her soft hand still caressed his cheek, keeping his face close to hers—and that his left leg was cramping up. He returned her soft smile, leaned back, reluctantly withdrawing from that delightful touch. He rearranged his long legs on the wing, grimaced a bit, then raised an eyebrow and let that trademark, lopsided grin spread across his features.

"You'll never know how many times I've dreamed of that, Dawn, wondering if it would ever happen."

"Really? I thought I was the only one who had those dreams!" she chuckled, a pleasant, tinkling sound. She eyed him closely, tipping her head to match the angle of his. "That was very nice, Howard Aster. Shall we go fly now?"

You lovely rascal! he thought. "We shall, milady." He closed her door, dropped to the ramp, and circled the airplane once again.

Settled in the SR22's left-front seat, Aster helped Erikson don a set of David Clark noise-canceling headphones and adjusted the boom microphone until it almost touched the right corner of her lips. He tried not to stare at those perfect, moist surfaces, relishing the memory of their touch.

Minutes later, Aster was strapped in, flipping switches and checking the SR22's avionics system as several flat-screens came to life. As he completed each item on a prestart checklist, he pressed a button, which checked a small box on one of the displays. Finally, he turned to Erikson and winked. "Ready to go?" he asked, his voice electronically altered through her headphones.

"Roger that, sir. Let's go!"

Aster's expression changed abruptly, and he glanced down to his left. He fished a vibrating handheld communicator from a belt clip, and shot an apologetic grimace at Erikson as he removed his headset. With hers still in place, she couldn't overhear the brief exchange, but her pilot's darkening features spoke volumes. Instinctively, she knew there would be no evening flight, and no quiet, private dinner with this wonderful man. Not tonight.

Erikson removed the headset as Aster returned the communicator to its holster-clip. He looked at her, wearing a pained expression. "Viking, I'm terribly sorry, but I've gotta get back to headquarters. . . ."

She shook her head, held an index finger against his lips, then hung her headphones on the right-hand sidestick control. A yellow-blond strand fell across her right eye, but it failed to hide a tear welling up. "Shh! Not a word!" she commanded. Stunned, he froze. She turned away, staring out the right door's window. He waited, watching her press a clenched fist against her lips, struggling for control. Confused, Aster flicked off the last few switches, then the master, leaving the instrument panel dark again. A heavy silence settled over the cockpit.

Finally, Erikson turned to face him, her eyes wet. In a barely audible, strained whisper, she choked, "Howard, when I was lying on the floor of the LAX terminal, with those horrible terrorists shooting and screaming, I was scared. Not scared that I might die, but scared of something I had no right to think of at the time. I was afraid I'd never see you again. After all those years, I was on my way to you . . . but I thought . . . I might never get here!"

She broke down, sobbing, head buried in both hands. Aster hesitated, unbuckled his lap belt, then reached over and did the same for Erikson, allowing her to twist in the seat. He wrapped a long arm around her shoulders and gently pulled her close. Though awkward in a confined cockpit, he managed to cradle her, letting the woman bury her face in his chest. A suggestion of perfume reached his nostrils, accentuating the growing ache in his heart, a pain that seemed to flow into Dawn Erikson, causing her to sob harder and longer.

The sky was darkening when she finally pulled herself free of his arms. Embarrassed, she tried to wipe her eyes clear, wordlessly accepting his white linen handkerchief. She finally glanced at him and tried to laugh, but the sound was a squeaky cross between laugh and strangled sob. "Good grief, Howard! You and Hans must be the only guys on Earth still carrying these things! You're *so* old-fashioned!" She again tried to laugh and blot her eyes without destroying their hint of eyeliner.

Aster half-grinned, took the handkerchief and gently wiped tears from her cheeks, carefully smoothing streaks of makeup. "Yeah, I'm old-fashioned," he admitted. "Mom made me carry a freshly ironed handkerchief to school every day, just so I'd be prepared to mop up some beautiful girl's tears. A hundred years later, it's finally needed. You all right?"

Her head bobbed the affirmative, but her lips were drawn tight, quivering. She was fighting the threat of another burst. "I'm so sorry . . . ," she started. This time, Aster held a fingertip against *her* lips.

"Shush, angel. Don't ever apologize for these feelings."

She took his hand and held it tightly between hers, looking deep into his eyes. A long breath settled her. "Howard, what scared the hell out of me in Los Angeles was this: I thought I'd never have an opportunity to say 'I love you.' I've loved you for a very long time; since we were in Qatar. But you were married, and I saw how much you adored Angie. There was no place for Viking in your life. I knew that. Now, you're free, and *I'm* married. But I gave my word, just as you did. I know there may never be a time for '*us*,' but it's important that you know how much I love you. And I always will. So . . . there you have it."

She might as well have slapped him, given the shock on his tanned, lined face, Erikson thought. *That* prompted a toss of her head and a laugh, a hearty, genuine laugh this time. "Whoa! Did I just scare the bejesus out of the sierra-hotel, fighter pilot General Aster?" she teased. Such an out-of-context remark broke the spell of seriousness that had enveloped the SR22's cockpit.

"Well, I . . . uhh . . ." Aster stammered, struggling to recover his emotional wits. He took a deep breath and tried again, "Damn it, Dawn! You'd drive a man to drink! Hell, I love you, too, but I had no idea. You never . . ."

"You assumed I was just another blond chick in a uniform? A wartime 'angel' looking for a man—*any* man—to keep her warm at night in the desert?" she challenged, tossing barbs that stung. But the smile said she was teasing.

"Hell, I . . . I don't know what I thought! You . . . It was impossible to tell whether you really cared or not, that's all." *Shit, was that ever adolescent!* he fumed silently, shaking his head in self-disgust.

"Yes, I really, truly cared. Unfortunately, I still care. I care far too much," she sighed. "But life is what it is, and our timing has really been a cruel pain in the ass, hasn't it?" She placed Aster's hand on his thigh and patted it firmly. "Now, a lot of people are waiting for the STRATCOM commander, and Viking is keeping him from responding to the call of duty. We'd better go."

Aster nodded, but leaned across the cockpit and kissed the attractive general's perfect, rose-shaded lips. Briefly, gently, but with authority.

"You're right, Viking. Frankly, our timing really sucks," he said, emotion tainting the words. He opened the left door and added, "My apologies for screwing up our evening. It appears that Chavez's VenezSat *is* a potential threat now."

Erikson had opened her door, as well, but swung back to Aster, all business. "It is? How?"

"The little bugger is on track to intercept one of Bigelow Aerospace's inflatable habitats in about two days."

"The Sundancer habitat?"

"Rog. And it has three people onboard."

She stared at him, her mind racing. "But why . . . ?"

"I was hoping you could tell me, Miz NSSI chancellor. At this moment, I have no idea why Chavez would drive a multimillion-dollar research satellite over to a private, manned habitat module, which is in a much lower orbit. But I don't have a good feeling about it."

STRATCOM HEADQUARTERS/COMMAND POST

"What do we have, Ted?" General Aster clipped, as he entered the STRATCOM command post.

"Nothing new, since I called you, sir," U.S. Navy Vice Admiral Ted Fraser, STRATCOM's deputy commander, answered, pointing at a wall display. "Fourteenth Air Force's conjuctive analyses confirmed that VenezSat is definitely on an intercept course with Bigelow Aerospace's Sundancer habitat. After an initial maneuvering burn, the bird's cross-track trajectory's been unchanged. Intentions unknown."

"How long until VenezSat reaches Sundancer, sir?" asked Major General Dawn Erikson, scanning the bustling room's mutiple wall-sized displays. This command post was the heart of STRATCOM operations, staffed around the clock, every day of the year.

" 'Bout two days, according to Fourteenth. My apologies for scrubbing your flight, Dawn, but every hour that goes by diminishes our options for dealing with this critter," Fraser explained in characteristic blunt, gruff fashion. Still in top physical condition at age fifty-four, Fraser was a living legend throughout the Navy SEAL community. Aster had seen the naval officer in action during the second Gulf war, and personally requested Fraser be given a third star and appointed as STRATCOM's deputy commander.

"Evening, sir, ma'am," Lieutenant General Dave Forester nodded to Aster and Erikson as he joined the group, carrying a sheaf of

BOYD printouts. "My apologies for disrupting your flight, as well, Viking." He was butt-dragging tired and, at the moment, hungry to boot. Referring to the printouts, Forester summarized the situation.

"Bigelow's Sundancer was launched a few months ago on a SpaceX Falcon Nine. Basically, the module's an orbiting blimp that weighs about ten tons and measures twenty-eight-point-five by twenty-point-six feet. Just a month before North Korea popped off its nuke, Bigelow lowered Sundancer's orbit and put three astronauts aboard for system-checkout purposes. The high-altitude nuke detonation scared the hell out of those guys, and the company's been scrambling to get 'em back on the ground. Bigelow's very concerned about those astros soaking up lethal does of radiation. As luck would have it, Sundancer was in direct line of sight to the blast, so those boys *were* exposed to heavy-duty prompt radiation."

"What's the status of a company rescue mission?" Aster asked.

"It's on hold, sir. SpaceX has a fuel-valve problem that's delayed a planned launch for at least a week," Forester said. "Bigelow's CEO called General Sawyer at Space Command, looking for help. Of course, Buzz told him the U.S. still doesn't have a quick-response military system that can launch on short order, so there's not much we can do to help those three astros."

"Anybody told Bigelow about VenezSat's projected intercept with Sundancer?" Aster asked.

"Negative, sir," Fraser answered. "That's the other reason I called you. There's no guiding policy, doctrine, or strategy in place to cover this situation. We have a manned, commercial spacecraft at risk of attack by a foreign entity. But how should the U.S. national security space community respond? Is that information classified? What's our responsibility? Nobody's been able to answer that."

"True," Aster agreed, "but General Erikson and her NSSI team have worked through that issue. During their briefing this morning, they introduced several viable options for handling VenezSat, too." He turned to Erikson and placed a hand on her shoulder. "Dawn, s'pose it's time to implement 'Operation Retrograde'?"

"With only two days until VenezSat reaches Sundancer, it may be our *only* option," she said.

"Then we'd better move out, folks," Aster directed. "Viking, how 'bout giving Ted, Dave, and the senior battle staff a rundown on Retrograde. Meanwhile, I'll put Speed Griffin to work, getting Blackstar ready to fly." Fraser and Forester exchanged puzzled glances, but said nothing.

STRATCOM HEADQUARTERS/COMMANDER'S OFFICE

Annie had left the office hours earlier, so the coffeepot was empty and clean. *Damn! No caffiene hit,* Aster grumped silently. He flipped a light switch on, slid into the high-backed desk chair, and thumb-scrolled through his handheld communicator's contact list. His call was answered on the first ring.

"Speed, Howard Aster here. You still in the office?"

"Roger, sir," Brigadier General Hank "Speed" Griffin replied. "With that Venezuelan satellite on the move, I've been expecting to hear from you."

"How the hell do you know about VenezSat?"

Griffin chuckled knowingly. "Test pilot intel network, sir. One of my 'deep black' brothers—we flew together at 'The Site' in Nevada—is on Space Command's senior staff. As soon as Fourteenth concluded VenezSat might be a threat to Sundancer, he called to check on Blackstar's status."

"Then you know why I'm bugging you," Aster clipped. "Get your space suit on, bud. We need Blackstar ASAP. Is it back in business yet?"

"The new canopy's been installed and pressure-tested. It's about ready to fly."

Aster outlined the situation as STRATCOM saw it, emphasizing that three Bigelow astronauts' lives could be on the line. He summarized NSSI's Operation Retrograde concept, as well as the dearth of space policy and doctrine at the command's disposal.

"Cripes, General! Retrograde is one helluva dicey operation! Is it even legal?"

"General Erikson's NSSI space lawyers seem to think Retrograde is technically legal, assuming VenezSat actually *is* a threat to Bigelow's three astros," Aster said, choosing his words carefully. "But we're definitely winging it. We're in uncharted territory here, Speed, literally flying blind on most of these issues. Hell, the legal eagles can't even agree on whether or not we're obligated to notify Bigelow that Sundancer and those three astros *could* be at risk. Would notification constitute a breach of security? I can't say. But if VenezSat rams that inflatable module and those Bigelow astros die, how the hell do we justify *not* notifying the company ahead of time?"

"I'm with you, sir," Griffin replied. "This is complicated and just begging for a rash of unintended consequences, though. Has Retrograde been vetted by your wargamers?"

"Viking Erikson's NSSI team participated in a limited-scope wargame today, exploring a number of scenarios, including Retrograde," Aster replied. "Bottom line is: We don't have many options, because three lives are involved here. If we were absolutely certain that those astronauts wouldn't be harmed, I'd sit tight and see what VenezSat does. But I don't have that luxury. So, get your go-fast suit on, Speed. I need that Blackstar bird in space ASAP."

"Rog, sir. Looks like the right call. Here's another tidbit to brighten your day," Griffin added. "My insider at NSA says the agency has intercepted comm between a gaggle of Russian and Venezuelan space geeks. The Ruskies were helping Chavez's people conduct a series of complex maneuvering burns, putting VenezSat on its new trajectory. Without the Russian help, that bird wouldn't have a hope of intercepting Sundancer, I'm sure."

Aster didn't reply, thinking. "So . . . why the hell would Chavez's bird target a Bigelow module? What's in it for either Venezuela or Russia?"

"Can't say for sure, but my *guess* is Chavez is trying to make some kind of big-splash statement," Griffin offered. "If he knocks out a high-profile U.S. commercial platform in orbit, he instantly joins the big leagues of spacefaring nations. It puts Venezuela right up there with Russia, the U.S., and China, the only nations that have demonstrated an antisatellite capability."

"Yep. Makes sense. And if he kills a few Americans in the process, it elevates Chavez's stature among all the anti-American, Third World nations he's courting, including a billion Muslims. *Damn!*" Aster exclaimed.

"That's my read on it," Griffin said. "As for me flying, no-can-do, sir. Your boss gave a thumbs-down on my flying the next Blackstar mission."

"The SecDef? Why would he ground you?"

"This is close-hold, sir . . . ," Griffin hesitated. "Secretary Hurlburt wants me to keep an eye on the National Security Advisor."

"What's that son-of-a-bitch Vandergrift up to now?" Aster growled. He spun his desk chair and propped both feet atop a credenza, staring at a dark Nebraska sky. Venus was shining brightly, just above the western horizon.

"He's laying low, at the moment. Seems the president tore him a new one, but Hurlburt's very concerned about Vandergrift's back-channel link to that Chinese agent, Nu. The SecDef doesn't trust our esteemed security chief, so I'm the Pentagon's designated Vandergrift-watcher."

"Christ! All we need, right now, is a flare-up with China." Aster sighed, rubbing his neck. He could feel the beginnings of a too-long-day headache taking hold. "But without Blackstar, Operation Retrograde is dead on arrival. I'll call the SecDef and . . ."

"Not yet, sir. We can still pull this off," Griffin interrupted. "We've shanghaied Alex Zeller, the Lockheed Martin contractor test pilot I told you about. 'Zulu' flew the XOV several times in the nineties. He's already taken up residency in the spaceplane simulator, prepping for an emergency satellite-reconstitution mission. Zulu's an outstanding pilot; he'll be good to go in short order." Griffin then outlined how Blackstar XOV would accomplish Operation Retrograde as an add-on to a mission Buzz Sawyer's people had set up. "I think we can launch that mission within the next forty-eight hours."

"Works for me," Aster said. "*Damned* good work, Speed! Get with Dave Forester, my ops chief, and let's make this happen." Hearing leather creaking, Aster twisted to see Dawn Erikson settling into one of the office's overstuffed chairs. He waved hello, dropped his feet, and pivoted the desk chair.

"Rog, sir," Griffin said. "I'll coordinate with the spooks to get Retrograde added on—assuming everybody cooperates." Aster caught the subtle hint that Griffin may need four-star top-cover, should some naysayer try to derail the Retrograde secondary mission.

"Holler if you need a blocker to run interference, Speed. Just make damned sure we take care of those Bigelow astros, copy?"

"Copy, sir. The XOV and SR-3 carrier are being mated as we speak," Griffin said.

"None too soon. And keep an eye on Vandergrift. That guy is flat-out dangerous." The two generals bantered a bit, trading classical fighter-pilot barbs, before signing off, Erikson noticed.

Big boys and big toys, she smiled. *They might wear stars, but fighter pilots never quite grow up. Thank God they don't!* At times like this, she was grateful that confident, capable, battle-seasoned officers like Howard Aster and Speed Griffin were hands-on, flying her nation through stormy skies. *Thank Heaven for old-fashioned patriots!*

Aster plopped into another overstuffed chair, legs askew, and shot Erikson a lopsided grin. "How'd Ted and the battle staff take to your Operation Retrograde?"

"About like you predicted," she laughed. "They're onboard, though. Admiral Fraser advocated a . . . let's just say a 'more-lethal' approach, but General Forester peeled him off the ceiling. Your ops director is an impressive guy."

Aster agreed, then leaned forward, elbows on knees. "Dawn, thanks again. For everything. Professionally *and* personally."

She nodded, but said nothing. Her lips tightened and Aster thought he saw tears welling up, but her eyes held his a long, pleasant minute. She finally stood, heaved a sigh, and said, "General Aster, could you spare a few minutes to drive a girl back to her quarters? I have a plane to catch at oh-dark-thirty tomorrow."

He unfolded from the chair, took her hand, and flipped the lights off.

ONBOARD BLACKSTAR XOV-2/60,000 FEET OVER
THE PACIFIC

Alex "Zulu" Zeller waited calmly, watching a digital countdown clock on the instrument panel. Strapped into the two-stage-to-orbit Blackstar system's XOV-2 spaceplane, his pressure suit and helmet sealed, he felt more than heard a throaty roar from the SR-3 carrier aircraft's engines. He could see very little outside his windscreen, because his XOV-2 was nestled under the "mothership's" belly, its nose covered by a smooth fairing. The huge, white-colored, XB-70-like carrier aircraft was accelerating.

"Two minutes to drop, Zulu." Steady and calm, the SR-3 copilot's voice reached Zeller through earphones built into his astronaut-style helmet.

"Gaspipe's good to go," he replied, using a cryptic call sign that controllers had given the spaceplane early in its flight test program. His comments were transmitted to a ground-based control room at Groom Lake, a secret Air Force facility in north-central Nevada.

A glance at the clock showed double-digit seconds remaining until release. Zeller completed the last items on a prelaunch checklist, then verified his gold-tinted helmet faceplate was secured in the down-and-locked position. He felt a mix of heightened senses, anticipation, and a familiar calmness. Years had passed since he'd flown the XOV-2, but other highly classified programs had taken him into low-Earth orbit in the meantime. He cracked a smile, wondering whether he'd ever be able to tell his wife and two sons that he'd logged hundreds of hours in space. Probably not, given a Mafia-like code of silence that ruled the closed, super-classified "black world" in which he worked.

"Mach three-point-zero; on-track. Launch in five, four, three, two, one, *release!*"

Zeller felt a thump of jaw-like clevices opening, freeing the sleek, titanium-gray spaceplane, which dropped rapidly, leaving him weightless. A blast of stark, high-altitude sunlight forced him to squint, despite being filtered through his gold-filmed visor. *One-thousand-one,* he counted, giving the XOV-2 time to fall well clear of the SR-3 mothership, which was banking sharply to his left. Zeller punched a button on the inside of the throttle in his left hand. A sharp kick confirmed the XOV's aerospike engines had ignited. Instant acceleration shoved his body deeper into a conformal seatback.

A glance at the engine display confirmed all four banks of "spikes" were cooking, fed by a stream of boron-based gel fuel and what little atmospheric oxygen could be sucked from the air at 60,000 feet. Later, in space, liquid oxygen stored in pressurized tanks within the XOV's fuselage would automatically flow to the engines, keeping the fire burning.

"Gaspipe's clear; accelerating to pick up climb schedule," he radioed, easing the throttle forward. As the spaceplane's speed increased, he thumbed a trim button on the right-hand sidestick controller, ensuring the aircraft remained in level flight.

"Gaspipe, mission control shows you in the green. Cleared to climb."

Zeller acknowledged with a double click of the radio-transmitter switch, jammed the throttle full forward, and smoothly pulled the control stick aft to pitch the vehicle's nose skyward. Eyes flicking between the instrument panel and the view outside his canopy, he noted the Earth's curved horizon disappear under the XOV's shark-shaped nose. Zeller locked the pitch angle at eighty degrees nose-up, trimming out stick pressures as the spaceplane accelerated almost straight up. G-forces, a measure of that acceleration, felt like a huge hand pressing on his chest, making it difficult to breathe normally. Mach numbers clicking by rapidly were testament to ever-increasing speed. Outside, a deep-purple sky gave way to the dark vastness of space.

"Gaspipe, on the zoom profile. Lookin' great, Zulu," a flight test engineer confirmed. On the ground, a team of experienced, hand-picked engineers monitored hundreds of parameters measured by a web of sensors throughout the XOV-2, their data continuously transmitted to sensitive antennas on the ground. A dozen sets of eyes tracked temperatures, pressures, fuel flow rates, and other critical parameters, ensuring the spaceplane's myriad systems were all functioning as intended.

Somewhere above the 100,000-foot altitude, speed readouts automatically switched from Mach number to feet-per-second, a subtle reminder that the spaceplane was now above most of Earth's life-giving atmosphere. Zulu Zeller was back in space.

"Control, how's inclination?" he snapped.

"Nominally thirty-five degrees. Hold what you've got. Lookin' good." Zeller bumped the trim button again, then relaxed his gloved grip on the sidestick. He was on-track, his flight trajectory forming a thirty-five-degree angle with the Earth's equator, rocketing to a predetermined orbit. Indeed, the tiny XOV symbol on his pilot's flight display was inching along a purple arc, confirmation that he was flying the spaceplane to precisely the right location. *That's what test pilots do, isn't it?* he mused absently. Still, it was personally satisfying to put a flying vehicle exactly where it was supposed to be in three-dimensional space. Damn few people could do that consistently.

"Gaspipe, Control: Twenty seconds to engine cutoff—*Mark!*"

Zeller acknowledged, then focused on keeping his spaceplane astride that purple arc. This was where he earned the big bucks. If the nose dipped a tad too much, his payload would wind up in a too-low orbit. It would soon slow to a critical velocity, fall back into the thick atmosphere, and burn up.

Two microsatellites nestled in a tightly packed "Q-bay" a few feet behind the XOV's cockpit were vital to the U.S. Navy's far-flung operations. Designed and built by the Naval Research Laboratory, "Arthur" and "Lancelot," the highly classified satellites' code names, were dual-mission prototype spacecraft crammed with cutting-edge technology.

Satisfied that he'd done everything humanly possible to fly XOV-2 to the correct point in space, Zeller engaged auto-nav, the spaceplane's autopilot system. It was now up to the navigation system's quad-redundant computers to coordinate with the aerospike engines hundreds of times per second, ensuring the XOV hit the targeted orbit.

"Gaspipe, Control shows engine cutoff—*now*. Confirm."

A sudden lack of acceleration threw Zeller forward against the seat's restraint straps. An eerie silence and peaceful sense of weightlessness greeted him, a welcome change from the persistent g-forces of sustained acceleration.

"Gaspipe confirms a clean engine cutoff. We on-track?" Zeller's multicolored flight display seemed to say "yes," but ground control had access to highly accurate tracking-radar data. He didn't.

"Roger, Gaspipe. Right where we wanted you. Congrats on another perfect insertion, Zulu."

"Nice to be back in the Blackstar business, guys," Zeller replied, grinning widely. *Frappin' fantastic! And they* pay *me to do this!* he marveled silently, stealing a moment to revel in the magic of drifting above a multicolored Earth, suspended in blackness. He turned his attention to the reason for being here: deploying two microsatellites. Time was *the* most-critical parameter now, and he had a lot to do. Particularly critical now, since he'd picked up a secondary mission.

Over the next few minutes, Zeller remotely activated the clamshell Q-bay doors, extended a trapeze-like structure, and gently released the Arthur microsatellite. A few subtle nudges of in-space flight controls fired tiny vernier rockets, which moved the space-plane well clear of Arthur. A payload team at a classified east-coast location first commanded the spacecraft to deploy two surprisingly small, folded solar panels, ensuring the bird would receive a vital trickle of electricity for months to come.

Both microsatellites were powered by a cutting-edge, top-secret "proton battery" developed by a small laboratory in Huntsville, Alabama. If it performed as advertised, the ultra-high-power-density battery would discharge very slowly, while powering on-board electro-optical and radar imaging systems and a vital communications-relay package. Solar panel-generated electricity would top up the battery's charge, ensuring onboard systems could operate in orbit for years.

Together, the two birds would restore essential command-and-control communication links with two Navy carrier groups patrolling the North Pacific. Comm with those forces had been sporadic, at best, since that North Korean nuke had exploded, silencing a swarm of U.S. communications satellites in low-Earth orbit, as well as a near-space, blimp-like HARVe comm-relay platform shadowing the carriers.

Thanks to the miracle of modern microelectronics and photonics, the two microsats could also acquire high-resolution, visible-light, infrared and synthetic-aperture radar images over politically sensitive regions of the globe. Granted, they could not stare continuously at one area, but their final, elliptical orbits would be optimized to give lengthy views of Southeast Asia and Eastern China. *The intel guys must be in dire need of eyes on North Korea and China,* Zeller surmised.

Working methodically, Zeller completed the deployment of Arthur and Lancelot ahead of schedule, a feat applauded by the control team at Groom Lake.

"Gaspipe, ready to initiate Operation Retrograde," the flight test engineer announced.

"Stand by, Control. Reorienting," Zeller said. He carefully entered several navigation parameters and cross-checked them. "Confirm auto-nav settings, Control." Normally, all in-space nav data would have been preloaded in the navigation systems' computers, but there had been no time to do so prior to flight. The last-minute, add-on Retrograde mission could only be accomplished by relying, once again, on a human being in space.

Zeller relished that thought. Somehow, it felt like he'd won a subtle victory in the inexorable, losing battle between humans and "cyber-borgs," as he called computer-centric systems. Yeah, 'borgs could do some things far better, definitely with greater precision, than a man, but 'borgs weren't as flexible. Human gray matter, the stuff between a human's ears, was still the most flexible, adaptable computer on the planet—or above it, in this case.

"Control, Gaspipe's in nominal position, 'spikes in the green, auto-nav armed. Ready to initiate cross-track burn."

"Control confirms positioning. Stand by for five-second burn; ignition on my call—Five, four, three, two, *burn!*"

Zeller again punched the throttle-mounted engine-start button. He felt and heard two banks of aerospike powerplants fire, silently praying that the auto-nav system would deliver the perfect combination of thrust and attitude. If something went awry, he and the XOV could wind up hopelessly adrift in the wrong orbit, potentially unable to safely reenter the Earth's atmosphere and fly back to Groom Lake. He shoved that thought aside and focused on the XOV icon drifting across his pilot's display, inching toward the new orbit's magenta line.

"Engines auto-cutoff," Zeller announced via radio.

"Gaspipe, confirmed. Stand by, Zulu. We're recomputing your new track." Zeller waited. He hated the waiting, knowing these next few minutes of cross-track drift spelled the difference between mission success or, worst case, new threats to his pink body.

"Gaspipe, Control confirms you're on-track. Nominal burn, nominal attitude. On-course to intercept VenezSat's orbit in two minutes—*Mark!*"

"Gaspipe copies. Great job, Control," Zeller replied, knowing his sense of relief was shared by those on the ground. *Thanks, Lord,* he added, to himself.

Zeller loved flying, and he was hopelessly addicted to the adrenaline rush that came with probing the boundaries of flight, the elite realm of test pilots. But he also cherished life. Since man had first taken to the skies, myriad test pilots had died "pushing the envelope," and he had no intention of joining their ranks. Zeller prided himself on using every tool and skill at his disposal to mitigate risk and minimize hazards. But, at a time like this, a pilot's fate was beyond his control. It rested with the vagaries of mechanical systems, electronic circuits, microprocessor calculations, and attention-to-detail diligence of some faceless software-code writer in a far-away cubicle. That's when Zeller and his fellow test pilots privately turned their fate over to a higher power. There weren't a lot of atheists in the flight-testing fraternity.

"Gaspipe, two miles from target. VenezSat should be at twelve o'clock, thirty degrees above you," the controller said, tension coloring his words.

Zeller searched the sky, slowly shifting his eyes left to right. *Gotcha!* "Control, tallyho; target in sight," he announced. "Ready to go manual."

"Stand by, Gaspipe. Auto-nav will get you a bit closer." The controller was subtly telling Zulu to be patient; let the 'borg do its job first, before turning control back to the pilot.

Zeller watched the slow-motion ballet unfold, as the XOV flew below Venezuela's "research satellite," the spaceplane's vernier rockets firing in quick bursts to bring the XOV onto a parallel flight path. He watched VenezSat drift overhead, maybe a hundred yards above his canopy, its solar panels glinting in the ultrabright sunshine. He wondered if the spacecraft's optics were tracking him, but the bird displayed no outward indication that it knew he was in the neighborhood.

"Gaspipe, you're cleared manual. Good luck, Zulu," the controller said tightly. It was now up to Zeller. Gently, he initially maneuvered his craft to a position ahead and below the target satellite. A few quick bursts of verniers raised his orbital trajectory, allowing the XOV to slowly narrow the gap with VenezSat. Using only his fingertips to avoid an unintentional roll input, Zeller cautiously pulled the control stick back, starting a slow pitch-up, back-flip maneuver, guiding the spaceplane's nose until it aligned with what would be considered the leading-edge surface of VenezSat's body. A few bursts of the XOV's aft verniers closed the separation to about thirty feet, drifting slowly.

The pilot mentally reviewed his task: essentially flying tail-first, he would guide the XOV until its nose contacted the satellite. A short-duration burst of longitudinal-axis thrust would serve to decrease the satellite's velocity, causing VenezSat to start dropping toward Earth. At some point, it would settle into a much lower, "retrograde" orbit. If calculations performed by Fourteenth Air Force's space wizards proved correct, Hugo Chavez's pride and joy spacecraft would not be able to boost itself back up to its former orbit. At that point, it would no longer pose a threat to Bigelow Aerospace's Sundancer habitat module and its three astronauts. Stuck in a lower orbit, VenezSat would be subject to Earth's upper atmosphere increasing drag on its solar panels and other surfaces, gradually slowing the spacecraft. Its trajectory would steadily grow steeper as velocity decreased, until the spacecraft fell into the atmosphere and burned on reentry.

Without warning, Zeller was blinded by a white-hot flash. For a millisecond, his brain registered a ragged-edged object flying directly at him. Then his world turned black, as debris smashed through the XOV's canopy, crushing the transparency with tremendous force.

"Gaspipe, Control! We've lost telemetry. Hold position! *Hold position!*"

The controller's rapid-fire plea was met with silence. Miles above the Pacific Ocean, VenezSat had inexplicably exploded. An onboard proximity fuze had sensed the presence of Zeller's spaceplane and detonated a compact cube of plastic explosive. In the vacuum of space, the explosion created no blast wave, but drove hot, extremely high-velocity debris through the XOV's canopy, fuselage, and fuel tanks. The pilot's faceplate was shattered instantly, shards slicing into exposed flesh. Without the protection of an intact pressure suit, Zeller's blood flash-boiled, mercifully killing the pilot immediately. The XOV's hull was ripped apart by searing hunks of metal, composite plates, and optical glass, forcing liquid oxygen to merge with tiny globules of boron-based fuel and triggering a secondary detonation. Pieces of the shattered spaceplane and VenezSat were propelled in an ever-expanding sphere, consigned to the growing band of space junk encircling the globe.

"Radar's showing *thousands* of new targets! Gaspipe's tracking beacon's disappeared!" an engineer shouted. Other mission engineers at Groom Lake's control facility chimed in, each reporting that XOV data traces had flatlined. Frantically, the mission controller repeatedly transmitted, "Gaspipe, Control! *Do you copy?*" He held his breath, straining to hear, but detected only a faint whisper of background noise on the dead communication link.

An Air Force colonel standing behind the lead controller finally laid a hand on the younger man's shoulder. "We've lost Gaspipe," he announced quietly over the mission-control intercom network. "That damned satellite blew up." Silence, shock, and disbelief hovered over the room. Nobody moved. Men and women continued to stare at uncaring workstation screens or each other, each struggling to grasp the magnitude of what had occurred in an instant of horror. The XOV spaceplane had been destroyed. Alex "Zulu" Zeller was dead.

# '11 COVERT STRIKE

WHITE HOUSE SITUATION ROOM

President Pierce Boyer listened intently as Brigadier General Hank "Speed" Griffin concluded his somber report about the on-orbit loss of the Blackstar XOV-2 and its pilot, then surveyed the Situation Room's other occupants. Defense chief T. J. Hurlburt, National Intelligence Director Herb Stollach, and Paul Vandergrift were unusually grim faced. Only the Secretary of State, "Mac" McAuley, revealed no discernable emotion, as he jotted notes in a leather-bound folder.

"Thank you, Speed," Boyer began solemnly. "Please convey my personal condolences to Alex Zeller's family. I understand the security constraints involved here, but it's absolutely essential that the pilot's loved ones know this office is aware of—and deeply grateful for—Mr. Zeller's courage and sacrifice."

"Certainly will, sir," Griffin replied. "I've been in touch with Zulu's wife, and have told her as much as we can about the 'accident.' She's taking his loss hard, Mr. President. Very hard."

The president nodded in sympathy, then took a deep breath. "Gentlemen, Hugo Chavez has committed a shocking, unprecedented act of war in space, a realm that, until recently, has been largely dedicated to peaceful coexistence. Venezuela's 'killersat' has taken the life of an American hero and destroyed an extremely valuable spacecraft. Only by His grace did Chavez fail to carry out a premeditated, murderous attack on a privately owned and operated space platform. Had he succeeded, Chavez would have killed *three* astronauts, and scored a major public relations coup among Third World nations."

Standing, Boyer leaned on the long conference table, glared at each person, in turn, then slammed a fist on the polished surface. "This despicable act *will—not—stand!*" the president shouted, his fist repeatedly striking the table for emphasis. He straightened slowly, crossed his arms, and started pacing, letting a strained silence hang.

"The question before us now, is: How do we respond?" Boyer finally asked rhetorically. "Since Zeller's flight was a covert, 'black' operation, using an American spaceplane that cannot be acknowledged openly, a diplomatic protest is not an option. That right, Mac?" McAuley agreed.

"Speed," Boyer continued, "as I understand it, the only Blackstar spaceplane in existence was the one Zeller was flying. Correct?"

"It was the only *man*-capable spaceplane, sir," Griffin qualified. "An XOV-1, a smaller, unmanned version, was used early in the Blackstar flight test program, but it hasn't flown since the early 1990s. It's not configured to carry a payload, either."

Boyer nodded. "Then another 'Rods from God' strike obviously is no longer an option. T.J., by tomorrow morning, I want a list of candidate measures for retaliation. Nothing that will start a shooting war with Venezuela or Russia, of course. Something that *will* hurt that mouthy SOB Chavez, though. I want to hit Venezuela *hard*, leaving no doubt that Chavez has incurred the wrath of America's might. But no U.S. fingerprints and no American casualties or prisoners. Any operation I authorize *will* meet the criteria of total deniability. No special forces on the ground and no overflights, understand? I will *not* have Chavez parading a beaten up pilot through the streets of Caracas, in front of a CNN camera! Understood?" Arms still folded, Boyer's gaze swept the room.

"Absolutely, Mr. President," Hurlburt replied. "The Joint Chiefs have already come up with strawman options, and several meet your criteria. Howard Aster's battle staff and STRATCOM wargamers are running the chiefs' candidates through a mini-game as we speak."

"What sort of candidates?" Boyer demanded.

Hurlburt hesitated, "Some . . . good ones, sir. The wargamers are looking at a spectrum of potential outcomes, political and military ramifications, and possible unintended consequences. Right now, I don't want to prejudice their deliberations with inputs from Washington. Let's wait until the 'gamers have completed their work, *then* we'll get into the act." Hurlburt pointedly addressed his last statement directly to Paul Vandergrift. The National Security Advisor rolled his eyes.

Boyer's chief of staff, Gil Vega, opened the Situation Room's door, drawing Boyer's attention, and jammed an index finger at his watch. The president raised two fingers in answer, and turned to Hurlburt. "T.J., it's imperative that your STRATCOM wargamers get those retaliation options to me as soon as absolutely possible. We have *got* to hit Chavez hard, and do it soon."

"Copy all, Mr. President," Hurlburt clipped. "We'll have something for you first thing tomorrow."

Boyer left the room, with Gil Vega in tow. While the others huddled to coordinate their next steps, Paul Vandergrift slipped from the room. He hurried to his office, told an assistant he could not be disturbed, and closed the room's heavy door. Tall, perpetually tanned, with perfectly styled, silver-gray hair, he was a striking figure, a man who had worked hard to attain his lofty White House position. But National Security Advisor was hardly the end of his Washington ascension. Paul Vandergrift had set his sights on the ultimate, highest office of the land. And he was on track to move into the Oval Office, perhaps very soon. For that reason, he tolerated the denigration, the put-downs, the butt-chewings from Pierce Boyer. Biting his tongue and biding his time now would pay off, eventually.

Vandergrift slid into an expensive leather desk chair. Rolling a thick Montblanc fountain pen between his palms, he silently reviewed his options. *Pierce is out of his element. He's losing it! Getting so worked up over that silly Venezuelan satellite incident! How pitiful,* he fumed silently.

Real leaders understood that losses were inevitable, the cost of boldly pursing worthy national goals. Indeed, lives were often the price of ultimately *winning* the complex chess match of international affairs.

Get over it, Pierce! Bigger, much more important issues than a spaceplane and pilot are facing the nation! Boyer could no longer be trusted to make the right decisions, the NSC director concluded.

As that insight crystallized, a tight-lipped smile spread across his features. Before Boyer completely fell apart, Paul Vandergrift would implement a daring, bold strategy, a plan that would literally save the United States and return it to superpower status.

He unlocked a desk drawer and retrieved a compact, tungsten-colored handheld communicator. The most advanced version in existence, his unit was one of a handful issued to top U.S. government officials. The devices had been designed and built under a classified, no-competition, National Security Agency contract. Consequently, they incorporated the highest level of encryption technology available.

Punching in a memorized number, he reflected on how he'd cleverly obtained this second communicator. His first had been "lost," while cruising with the president and his Australian counterpart. The White House comm-security register duly reflected that the communicator had "accidentally fallen into a deep-water lake." It had been difficult, but Paul Vandergrift had managed to put that "lost" device in the hands of a very special contact, Feng Bao Nu. The U.S. National Security Advisor heard that special contact answer on the third ring.

"Yes, my friend?" A Chinese accent was unmistakable.

"I trust the dragon is prepared?" Vandergrift asked.

"Yes. And the eagle?"

"The eagle will fly within forty-eight hours."

"The dragon will awaken," Nu said, then broke the connection, ensuring the call could not be traced or intercepted.

However, miles to the northeast of Vandergrift's office, a powerful National Security Agency computer automatically captured the short burst of voice traffic from a specific White House-registered handheld communicator and immediately activated a decoding algorithm. Soon, an alert, followed by a clear-text version of the Vandergrift-Nu conversation, appeared on Colonel Brian "Rattler" Rich's computer screen. Rich swore softly, grabbed his own communicator, and dialed another White House-registered device.

Within thirty minutes of Vandergrift's call, a small number of military officers and the Secretary of Defense knew yet another Paul Vandergrift "black op" was unfolding.

"MUSKET" FLIGHT/30,000 FEET ABOVE CARIBBEAN SEA

"Stand by to disengage," the air-refueling "boomer's" voice warned. Captain Jason "Pepper" Malloy waited until the winged "pipe" of the boom lifted and retracted, its valved, circular tip coming into view near the bubble canopy of his F-22 Raptor. He tugged the fighter's throttle aft, eyeing the tanker's boom as his Raptor fell behind and well below the KC-10 flying gas station. A nudge of the sidestick controller in his right hand brought the F-22 to a perch above and to the right of the huge green-painted tanker's tail.

Malloy took a deep breath, which triggered a whooshing sound in his helmet's earphones. The pilot rolled his neck to ease tense muscles. Although he had a thousand hours of flight time, aerial refueling remained one of the fighter pilot's most demanding, high-concentration, high-workload tasks.

Malloy completed the post-refueling checklist, verifying the jet's air-refueling door was closed. He watched the final bird in his five-fighter flight, a U.S. Navy EA-18G "Growler" electronic attack aircraft, approach an oversized shuttlecock-looking "basket" attached to the end of long hose trailing from the KC-10's right wing. The Growler's arm-like refueling probe deployed, angling away from the fuselage. The pilot made short work of inserting the probe into the basket, causing the hose to droop. Malloy wondered, again, why the Air Force and Navy had never settled on a common air-refueling method. Air Force fighters were refueled by a long, centerline boom that the KC-10's operator or "boomer" literally flew into a receptacle in the receiving aircraft's skin. Navy fighters extended a bent-arm probe, then flew the device into a basket.

As the Growler topped its tanks, Malloy had to admit that he was damn glad the Navy's newest electronic warfare bird had been assigned to his mission. His impression of its capabilities had changed considerably, during intense mission-planning sessions at Strategic Command headquarters. The Growler was advertised as the replacement for EA-6B Prowlers, the U.S. military's jointly operated electronic warfare platform. The Prowler's jamming had blanked many enemy radar screens over its long and storied career, protecting hundreds of strike forces penetrating deep into bad-guy territory.

But the EA-18G Growler could do far more than simply jam an adversary's radar, air defense, and communications networks. The new bird was a true twenty-first-century electronic attack platform. Even though the Growler was still being flight tested, the Pentagon's star-studded powers had decided it was time to "bloody" the aircraft in a no-shit combat operation.

Waiting, the pilot mentally replayed the 1st Fighter Wing commander's declaration that he, Pepper Malloy, would be leading a highly classified, first-of-its-kind strike mission. He still couldn't believe his incredibly good fortune. That *he* had been tapped to be this mission's flight leader was proof-positive that a Higher Power had a soft spot in his heart for fighter pilots. Especially since Malloy and his wingman, Captain K. B. "Shark" Fisher, were still under the cloud of an Air Force investigation into a mission-from-hell a year earlier.

The two F-22 pilots had supported a deep-black, off-the-books, Army Special Operations Forces hit against a terrorist cell in Algeria. Through a horrific twist of high-tech, perverse fate, four damaged GPS satellites had been responsible for Malloy's two small-diameter, GPS-guided bombs landing a hundred meters from their intended target, blowing up the American SOF team, not the bad guys. Yeah, he and Fisher had recovered and nailed the terrorists before they got away, but Malloy would never forget the vision of that horrible fireball in the desert. And he had to live with the knowledge that *his* bombs had killed some of the nation's finest shadow warriors.

"I can't imagine who in Washington you two yay-hoos could possibly know," the wing commander had drawled, "but the Secretary of Defense himself strongly 'suggested' that you, Malloy, command this mission." The officer had shaken his head in disbelief, adding,

"Most political appointees would have hung you two after that goat-rope in Algeria. You damn sure lucked out by having an old Army general-turned-SecDef holding your career-gonads in *his* gnarly fists. SecDef Hurlburt said, 'Get those two pilots back in the saddle. That tragedy in Algeria wasn't their fault.' When I was a captain, the SecDef in those days would have sacrificed a dozen fighter pilots to save his own ass. You boys are damned lucky, I tell ya."

The days following that session had been a mind-numbing whirl-wind. Malloy, Shark Fisher, two Navy F-35 Lightning II test pilots from the Naval Air Station Patuxent River flight test center in Mary-land, and a two-person Navy EA-18G crew had been flown to Offutt AFB, Nebraska. There, Strategic Command's battle staff, augmented by experienced Air Force and Navy fighter-squadron planners, had worked with Malloy's team to plan the most high-tech, amazing combat strike he could have imagined. Now, it was his job to make it happen.

Its fuel tanks full, the Growler eased away from the refueling bas-ket, dipped a wing, sliced beneath the F-22, and reappeared to his right, joining three other fighters stacked off Malloy's right wing. The captain waited until all five aircraft—two F-22s, two F-35s, and one EA-18G—were stabilized in right-echelon, the nose of each successive aircraft slightly aft of the fighter nose to its left.

Malloy punched a key adjacent to one of his Raptor's color displays. Instantly, a curt, encrypted message was fired by a narrow-beam, low-probability-of-intercept datalink to the other aircraft. The message, *BUTTON 4*, appeared on the other pilots' cockpit displays. Malloy waited a long beat, then thumbed a radio-transmit switch on the in-side of his right throttle's handle.

"Musket flight, check," he clipped. In quick succession, the other pilots responded.

"Two."

"Three."

"Four."

"Five."

Malloy said, "Go mission profile." Four radio-mic clicks an-swered. If the mission proceeded as planned, that brief exchange would be the last voice communications necessary, until Musket flight rejoined the tanker on its way back to Florida. Normally, the flight would be tracked by a boatload of senior officers in the STRATCOM command post, but not today. That North Korean nuclear blast had taken a serious toll on America's national security satellite fleet, cutting the "reach-back" lifeline that had become the norm for U.S. military operations across the globe. Without those spacecraft and that comm lifeline, Musket flight was on its own. And that wasn't all bad, the Musket mission commander decided.

Malloy's right hand nudged the Raptor's sidestick inboard, standing the fighter on its left wing. He pulled the stick aft, until the nose pointed south, then rolled wings-level, but in a nose-up cruise-climb. The other jets followed. A quick glance at the navigation cockpit display showed they were less than two hundred nautical miles from the shores of Venezuela.

Show time! Malloy smiled.

MUSKET FLIGHT/OVER THE CARIBBEAN SEA

Pepper Malloy pulled the throttle back and lowered his fighter's nose, smoothly leveling the jet at 65,000 feet. He fiddled with the radar controls, until a clear, photo-quality synthetic aperture radar (SAR) image appeared on a large-format flat screen embedded in the F-22 instrument panel. Far to his right, Malloy could see Shark Fisher's dark-gray Raptor flying a parallel track, pointed at a huge oil refinery complex. A key tap verified the port area on Venezuela's coast was one hundred twelve nautical miles away.

Malloy switched the AESA, or "active electronically scanned array," radar's mode to search, automatically commanding invisible beams to sweep the sky, looking for Venezuelan fighters that might be streaking toward Musket flight. None—so far.

A cryptic data link message from Shark appeared on Malloy's screen: *HOP STLTH SHT WRKS.* Malloy chuckled. He, too, hoped and prayed the Raptor's high-priced "stealth," or low-observable shape and special materials would prevent ground-based air-defense radars from detecting the American fighters. He took comfort in knowing the weapons bay below his cockpit carried four air-to-air missiles: two medium-range, radar-guided AIM-120 AMRAAMs and two short-range, heat-seeking AIM-9X Sidewinders. Should any of Chavez's fighter pilots feel lucky, they'd quickly discover they were no match for the Raptor, and the foreign pilots would be making silk-chute landings—*if* they survived a U.S. missile up their tailpipes.

Protecting Musket flight from attackers was Shark's and his primary job, but he really didn't expect to be engaged by Venezuelan fighters. Myriad flight tests, exercises, and training missions had confirmed that the F-22's and F-35's "stealth shit" worked extremely well. The EA-18G Growler, encumbered by several electronic-warfare pods and underwing missiles, was hardly stealthy, but it would be positioned farther away from shore than the Raptors and Lightning IIs, at an altitude optimized for its role.

A burst-scan of Malloy's AESA radar detected a sixth aircraft that was critical to Musket flight's mission, an RQ-4 Global Hawk. The bulbous-nose unmanned aircraft had been on-station at 60,000 feet, orbiting for hours, its sensors and radar probing the Venezuelan air defense system, as well as the refinery complex. A datalink query of the Global Hawk was answered with a flood of data. Within minutes, all five Musket aircraft had received enough information for powerful onboard algorithms to define an "electronic order of battle."

A few thousand feet below Malloy, two Navy F-35C Joint Strike Fighters cruised, their AESA radars already probing computers, electronic switches, and other key network nodes along Venezuela's shoreline and on the surrounding hills. The Global Hawk had soaked up spurious signals, identifying the location of air defense search-and-track radars, ground-to-air missile batteries, and key communication nodes. Now, those downloaded data greatly simplified the F-35 and F-22 crews' final tasks: identifying their targets.

The EA-18G Growler crew also took advantage of the Global Hawk data-dump, selecting vulnerable nodes in the air defense network. In the Growler's rear cockpit, weapons systems officer Lieutenant Jennifer "Fearless" Fernandez tapped several keys, telling the aircraft's powerful computers what subtle commands should be riding on their AESA radar's beam.

"Okay, Cube, targets are loaded," she said, her tone all business. "Standing by."

"Rog. Shoot some lines to Pepper. Tell him we're cocked and ready to conduct the symphony," the pilot answered. The compact, muscular Lieutenant Commander Joshua "Cube" Stone and his WSO, Fernandez, had wrung out the Growler's mind-boggling systems and software through hours of tedious, methodical flight tests. They knew how to manipulate the electronic magic of ALQ-99 jamming pods under each wing, as well as perform unbelievable "information warfare" ballets via the AESA radar tucked inside the Growler's nose cone. But they'd never fired their airplane's electronic systems "in anger," targeting a real-world adversary. Both officers were anxious, chomping at the bit to demonstrate their bird's formidable electronic attack capabilities.

Pepper Malloy saw Fernandez's brief datalinked message appear on his screen. *All right! Let's rock!* he enthused silently. He fired a preformatted message to the other four aircraft, unleashing "Operation Tron." Hugo Chavez had no idea how his routine day was about to become a mysterious nightmare unlike any he could have imagined.

Digital pulses fired from the EA-18G Growler's AESA antenna flashed across the sky at light-speed, slipping undetected into Venezuela's expensive, Russian-built "integrated air defense system," or IADS. The few radar pulses reflected from five American aircraft, particularly the Growler, appeared briefly on Venezuelan defense forces radar screens, then disappeared, unnoticed by operators. The Growler's covert "information operations" commands essentially told the digital blips created by the fighters to vanish, flushed from air defense system computers entirely.

Unknown to Venezuela's bored radar operators, their sophisticated systems were under serious electronic attack. Without the Growler's subtle e-trickery, the ground radars probably would have detected the EA-18G or the stealthier aircraft, prompting Venezuelan fighter pilots to run for their interceptors, and Russian-made surface-to-air missiles, or SAMs, to start searching the skies for intruders.

Tactics employed by the Growler were advanced, better-honed versions of an information-operation campaign Israel had waged against Syria four years earlier. The stunning mission allowed Israeli fighters

to slip through Syria's vaunted air defense web, undetected, and bomb
a suspected nuclear plant. The Syrians never saw the aircraft, because
their IADS computers had been told to ignore radar returns from the
fighters. Although none of Israel's F-15s and F-16s were "stealthy,"
their radar returns never appeared on Syrian 'scopes, never triggered
alarms that would have sent deadly missiles into the sky, aimed at
the fighters. On that day in 2007, a new benchmark of warfare was
established.

The F-22s and F-35s were miles and several thousand feet apart,
but their AESA radars were communicating, coordinating actions.
Software algorithms sent commands to hundreds of tiny transmit/
receive modules that blanketed the circular faces of AESA antennas, electronically steering agile radar beams at almost the speed of
light. Each tiny T/R module acted like a small, individual radar,
but the beams combined, focusing their power on faraway targets.
By performing thousands of computations per second, AESA computers adjusted each aircraft's radar signals, until all four beams—
two from the F-22s, two from the F-35s—converged on a key control
node, the huge oil refinery's control room.

In a flash, pencil eraser-sized microchips in dozens of computers
and control circuits were subjected to far more electrical current
than they could tolerate. The AESA radar beams comprised relatively moderate electromagnetic pulses, but powerful enough to
briefly induce high currents in the refinery's control system wiring
and circuits. Many of the targeted electronic components literally
fried, as if they'd been subjected to the EMP from a small nuclear
weapon. Although fired by AESA radars, the pulses had the same
impact on delicate electronic chips.

Computer microprocessors, before they failed, typically spewed
streams of spurious electrons, which were interpreted by other circuits as confusing, yet sometimes valid commands. As a result, chaos
spread rapidly throughout the oil refinery complex. Computer-
controlled valves slammed closed, causing fluid "hammer shocks" to
ripple through pipelines and into tanks. Some lines ruptured, spilling
crude oil, partially refined jet fuel, and raw gasoline.

Others opened, sending floods of liquid into lines and tanks, often in the wrong direction. Fail-safe systems automatically at-

tempted to limit the damage by commanding safety and check valves to open or close, but with only partial success. In less than a minute, the massive refinery was rendered inoperative. Physical damage was minimal, and spills would be contained fairly rapidly. But Venezuela's primary refinery and petroleum-shipping port would be out of business for many months.

Cruising at 65,000 feet over the tranquil, blue Caribbean, Pepper Malloy had no sense of the catastrophe Musket flight had inflicted on Hugo Chavez's refinery and port. After an allotted period, AESA radars on the F-22s and F-35s automatically reverted to other modes, their combined-beam, fused-energy mission complete. Only a few seconds of the right signals, at the right power levels, were required to damage sensitive microcircuits, even if the chips were miles away, inside buildings and computer chassis.

Malloy punched another key, firing a single datalink order to the rest of his flight: *MSN CMPLT—EGRESS.* Pilots of two single-seat F-22s and two F-35s adjusted their cockpit controls, returning radars to low-probability-of-intercept search modes, then started shallow-bank turns back to the north. Each man aimed his fighter at an imaginary spot in the sky identified by a set of GPS coordinates, the Musket flight rejoin point.

Separately, Stone and Fernandez closely monitored complex, data-rich flat screens in the Growler, ensuring no Venezuelan radar could detect the five American fighters.

"Hey, Cube! Did we outfox Mr. Chavez and his incredible, smartsy IADS or *what*? Ooooaaah! Yes!" Fernandez hooted from the rear cockpit, via intercom. "Mission complete, oh-courageous-pilot dude! Let's hightail it. Beat feet. Skedaddle!"

"Cool it, Fearless!" Stone ordered, his fingers flying across the front cockpit's displays and controls. "Soon as you sanitize the pods, I'll get the hell out of Dodge." It was a familiar airborne exchange the two had adopted, during their hours of flight-testing time together. Fearless played the carefree, devil-may-care unprofessional, while Cube, the senior officer and aircraft commander, evoked a serious, business-only persona. Once on the ground, the roles seemed to reverse. Tall and big-boned, Fernandez would never be mistaken for a former college cheerleader. Plain and pale-skinned, she was relatively quiet and studious—and a brilliant electrical engineer. Stone, on the other hand, had to stretch to make five feet five inches, but he practically lived in the gym. He could still masquerade as the Naval Academy champion gymnast he had been a decade earlier.

Stone reflected on those differences, waiting for Fernandez to secure the powerful jamming/electronic warfare pods under the Growler's wings and on its wingtips. Yeah, he and Fearless made an odd couple, but everybody in the Pax River EW test squadron would concede they constituted *the* best, smartest, and most effective Growler crew on Earth.

"Good to go, Cube. Light the fire and herd this puppy home!" Fernandez called. Stone pushed the EA-18G's twin throttles forward to full military power and pulled the nose skyward. A sharp stick move to the right, then back to center banked the Growler into a steep-angle, climbing turn. He rolled out on the correct heading, still climbing, and kept the throttles against the milpower detent.

In minutes, he was sliding into position off an F-35's right wingtip. The Lightning II's Navy test pilot flashed an enthusiastic, vertical fist pump. Stone glanced in the narrow rearview mirror on the front cockpit's canopy bow, as Fernandez enthusiastically returned the salute.

She's stoked, he smiled. Rightly so. The fair lady had just e-attacked one of the most formidable air defense systems in the hemisphere. Her skill and smarts were responsible for tricking Mr. Chavez's crack IADS radar crews, keeping them clueless and silent. The Venezuelans never knew Musket flight had been in the neighborhood. And nobody in Chavez-land suspected who or what had functionally destroyed Venezuela's primary oil refinery and petroleum-shipping port.

STRATCOM HEADQUARTERS/COMMAND POST BATTLE CAB

"Sir! Burst message from Musket flight!" An Air Force major at a milsat communications console waved the general over.

"What's up?" Dave Forester asked, leaning over the officer's shoulder, squinting at the color display. The major pointed at a terse, cryptic message: *MSN CMPLT. MSKT EGRS W/ 5. CD 1. . . .*

"All right!" Forester whispered. "Successful mission; all five jets outbound, all in good shape. That it?"

"That's all we received, sir. Damn satellite's transponders are on their last legs. It's a wonder we got that much through."

Forester slapped the major's shoulder, then stepped away and speed-dialed a number on his handheld communicator.

"Aster. What's up, Dave?"

"Musket's outbound, sir. We picked up part of a message through a WGS satellite," Forester reported.

"*Damn!* Maybe those kids did it!" Aster exclaimed. Forester could hear the four-star slam a fist on his oak desk. "As soon as Musket flight hits the ground in Florida, give me a yell and come up here. The debriefing will be teleconferenced to my office. Nice work, Dave. Kudos to your people!"

Forester signed off and flipped the device's cover down. He indulged in a moment of unbridled self-satisfaction, the afterglow of a tough mission successfully completed. It was rare that STRATCOM assumed full responsibility for planning and conducting a real-world tasking like Operation Tron. But the president's order that the mission be covert and close-hold dictated very few people be brought into a tight circle of those having a need-to-know. It was a stretch, but SecDef Hurlburt had decreed Tron was a "long-range global strike" and a clear-cut "information operations/electronic attack" mission, which tossed it cleanly into STRATCOM's job-jar. And by initial indications, they'd pulled it off.

Not bad for an Army helmet-head and a bunch of kids! he decided.

FOX NEWS TV REPORTER/ON-SCREEN TAG LINE:
"CARACAS"

"Shep, a massive computer meltdown paralyzed Venezuela's primary oil refinery and petroleum ocean-shipping terminal today, trapping hundreds of thousands of gallons in damaged pipelines and tanks. A refinery official tells Fox News that a freak electrical surge destroyed delicate chips in hundreds of computers and computer-controlled 'valve trees.' Cause of the power surge is under investigation.

"Because the refinery's network of controls was damaged, crude oil, gasoline, and other products were routed to the wrong places. For example, jet fuel ready to ship to the U.S. and other world markets was routed to storage tanks filled with sludge and waste products. Shep, we're told that millions of dollars of already-refined products were contaminated or destroyed this way.

"Details are sketchy, but the official said this massive complex you see behind me will probably be shut down for several months, while engineers and technicians repair the mess. Amazingly, there was no fire or explosion. For that reason, the official said investigators currently do *not* believe this was a terrorist attack.

"Venezuela's president, Hugo Chavez, was outraged by the accident, vowing to take swift action against refinery and port managers. He declined to speculate about the potential impacts of this disaster on Venezuela's economy, but a government spokesman claimed the suspension of oil shipments from here would have a, quote, 'profound, immediate, and negative effect,' on the nation's financial state.

"It's not clear how much of Venezuela's government largesse is funded by oil revenues, but it's substantial. Many of the president's social programs and international philanthropy is underwritten by proceeds from the oil industry. Dozens of companies were nationalized in recent years, pouring billions of dollars into Chavez's coffers—and driving big-name companies like Shell and ExxonMobil from the country.

"Fox News will be monitoring the fallout from this disaster. Breck Hedden in Caracas, Venezuela. Back to you, Shep."

12 THE GAMBIT

WASHINGTON, D.C.

Admiral Stanton Lee awoke to a smooth landing at Washington's Reagan National airport. He'd never liked the long drive from Andrews Air Force Base into the District, so he welcomed the courtesy of a VIP's National Airport delivery and being met by a chauffered SUV. Lee noted that Washington's teeming traffic had become as ubiquitous as monuments.

How do these people do it? Incessant traffic, gray skies, either biting cold or ghastly hot and humid, never seeing your family. And those mind-numbing staff jobs . . . If you're working in the Pentagon, you might as well be deployed.

He turned to the upcoming White House meeting. The president had been surprisingly receptive to Lee's frank comments about China, during that last videoconference. Then came the unexpected message from Zhang. *Serendipity can be fateful, and Zhang's note definitely changes the calculus with this president. Hope I'm up to this.* He sighed. *So much at stake . . .*

Despite traffic snarls and tight gate security, Lee's SUV finally arrived at the stately first-residence's portico, where a Marine Corps guard opened the vehicle's door. Lee exited and walked ramrod straight as he passed several security posts. The Oval Office appeared serene this morning. Fresh flowers accentuated the sparsely decorated room with a pleasant, barely discernable aroma.

The group Lee joined wasn't the intimate gathering he had expected, however. The president sat behind a rather small desk, flanked by Gil Vega, the chief of staff. Paul Vandergrift was present—but, of course, the National Security Advisor would about have to be—seated slightly behind and to Boyer's left. Lee proceeded to the only open seat, a straight-backed chair positioned directly in front of the president, the desk between them.

The retired admiral shook hands with SecDef T. J. Hulburt, who sat to Lee's left, and Secretary of State Mac McAuley to his right.

The president spoke first, skipping the pleasantries. "Admiral, you waxed somewhat prophetic when we last spoke. Since then, it seems the Chinese have become rather oblique about certain things."

Lee nodded, but remained silent. Boyer lifted a sheet of paper from his desk, holding it aloft.

"I am herewith reinstating your diplomatic status, and sending you back to China. I'm appointing you 'Ambassador-at-Large' for the purpose of engaging Chinese leaders in an attempt to avoid hostilities. That is your focus. That is your mission. You are to report directly to this office, and the gentlemen here all understand why I want a direct link. I expect you to coordinate your observations or recommendations with nobody but me. I also want the Chinese to understand they have a real-time, direct line to me, via you. Is that clear?" He glanced around, satisfied by a wave of bobbing heads.

"There's at least one person who is currently in the Chinese Embassy who knows about this initiative. A diplomatic pouch received last night contained an offer for a meeting after you arrive in Beijing. Actually it was worded more like a *demand* than an offer," Boyer corrected.

"Interesting. And his name, Mr. President?" Lee queried.

"Feng Bao Nu. We don't know much about him."

"I know him," Lee said slowly, choosing his words carefully. "He was a Division Chief in their Ministry of State Security. I suspect he is still with the ministry in some capacity."

Vandergrift couldn't resist, "Oh, that's great! A damn spy!" To Lee, the remark seemed contrived.

"After a fashion, but grown differently than ours," Lee responded. "The Chinese minimize the scientific method of intelligence-gathering—and analysis, which is so prevalent in our intel communities. They opt for more of a humanistic approach. The idea is to gain a better understanding of the social fabric of their intelligence target, and to be able to think as their adversary thinks. As you might imagine, most of their resources are directed toward us, at the moment." He spent several minutes describing critical, but subtle, cultural differences, which helped the president understand Chinese thought processes. The retired admiral sensed he was connecting with Boyer, who listened intently.

"Nu is equivalent to a scholar of the Taoist way. His knowledge of the Western world, particularly the United States, makes him a most-valuable asset in the ministry."

"Why does Nu want a meeting?" Hurlburt asked.

"I can't say with certainty," Lee shrugged. "There are at least two possibilities. First, the Chinese might want to make sure we understand their interpretation of the current space situation, underscoring how critical 'space situational awareness' is from their perspective. This is somewhat borne out by Dr. Zhang's note. Zhang is an international scholar, highly respected by China's leadership, particularly the moderates. And Zhang knows the degree of my sensitivity to Taoist ways.

"In contrast, Nu comes on the scene amidst secrecy. Nu and I disagreed sharply over an issue of U.S. policy: Which is it? Engagement, or containment?"

Lee continued. "The second possibility finds China's leadership split over what to make of our current geopolitical troubles, and what actions they should take. Let's say that hard-liners discovered the Zhang connection, and, for sake of argument, that they are now countering through the Nu connection. Nu will take full measure of me in any negotiations, and he will weigh his assessment against his knowledge of my ways. He will look to see whether I behave in a subtle manner, or more direct and headstrong, complete with ultimatums. If it's the latter, he will dismiss me as a red herring, someone simply doing the bidding of certain elements here in Washington."

"Well, let me make the 'elements' absolutely clear," Boyer exclaimed. "You represent *me*, Admiral. The President of the United States. They won't misunderstand that!"

"Yes, sir. And with respect, Taoism counts on that very characteristic in potential opponents: establishing the paradox of fullness and emptiness; winning without fighting. We in the West struggle with that very ambiguity. And yet, such ambiguity is China's science of psychology."

Vandergrift scoffed, "If we spent as much time planning intelligent moves as we did talking, we'd be a lot better prepared. Mr. President, we have *far* more pressing matters before us."

"You've given us much to digest concerning our approach to dealing with the Chinese, Admiral," Boyer said, ignoring Vandergrift. "Frankly, I don't know how much of it to accept. On the other hand, I'm not sure we have many choices here. I *am* curious about one thing, though. This 'shoes of the fisherman' comment. What's that all about?"

"The Chinese, as usual, have done their homework," Lee smiled. "Neither Zhang nor Nu were in my circle of Chinese associates, when I was the ambassador. One of our mutual colleagues may have exaggerated his characterization of me, imbuing yours truly with the qualities of Morris West's fictional protagonist."

Hurlburt nodded, recognition dawning. "Ah, yes! The Russian who leaves the gulag and becomes Pope. He staves off World War III by pledging the Vatican's wealth to feed a starving, desperate China. *Shoes of the Fisherman*. Great movie."

Vandergrift's sarcasm overflowed again. "If this don't beat all. Admiral, diplomat, Zen master and now 'Pope'? Look, all this crap about Too, Tie . . ."

"Taoism, Paul," interjected Hurlburt, his barely suppressed disgust with the NSC director evident.

"Yeah, whatever. But let's not forget about our own intelligence estimates and State's documented concerns. China is a human rights abuser, a WMD proliferator, an unreasonable trading partner that flouts World Trade Organization protocols, pirates our intellectual property, hacks our computers, and floods Silicon Valley with corporate spies.

"And it's now a military threat that could close off the Western Pacific," he added. "Those are hard, cold facts! I say we deal with those, and forget about what Confucius says. And I do *not* endorse the idea of Admiral Lee meeting with this Nu fellow."

Lee sat quietly. Vandergrift was a one-dimensional Beltway infighter, hardly up to thinking through the myriad elements of complex, difficult issues, let alone the vagaries of international relations.

Lee leaned slightly forward in his chair and held Boyer's gaze. "Mr. President, China's actions in many areas are not beyond reproach, and I don't mean to imply the Chinese have some mystic quality that we should defer to. I *am* trying to share with you how the Chinese think, and how we should consider their thought processes as we develop our own strategies," Lee stressed. He paused for effect.

"We should view the emerging crisis as an opportunity for integration. Maybe even a partnership, a strong partnership that can be sustained long after the current fires are extinguished."

The President eyed the retired naval officer and diplomat with respect. He then rose, circled the desk, and leaned against its front edge, close to Lee.

"I'll think about your comments, Admiral. My team here will brief you on the fine points of our position, as it's been developed thus far. I must say, sir, this has been a most interesting session. Most interesting."

Boyer extended a hand, flashing his best campaign smile.

STRATCOM HEADQUARTERS/COMMANDER'S OFFICE

As General Howard Aster headed for his office, he glanced at a placard. He'd seen it dozens of times, but its meaning now seemed stark and pointedly meaningful:

> *Act after having made assessments.*
> *The one who first knows the*

measure of far and near wins.
This is the rule of armed struggle.
Sun Tzu

Aster anxiously awaited Admiral Lee's report, following the trip to Washington. Whatever had unfolded during Lee's meetings with the president and his security team would affect not only the current wargame, but also the many issues STRATCOM was addressing. For one, the 'gamers would be taking up a highly sensitive diplomatic assessment involving China, and Lee was *the* China expert. The retired admiral's absence already was affecting that side-wargame's progress.

Lee arrived looking weary, after a whirlwind of multiple meetings. Aster met the retired naval officer at the outer-office door. "Welcome back, Admiral. Appreciate you debriefing me, before getting back to the 'gaming center."

"My pleasure, General."

"Look, I can't bring myself to call you Stanton," Aster said, waving Lee to one of the office's overstuffed chairs. "Just doesn't fit. I'd appreciate it if you'd call me Howard, though."

Admiral Lee accepted the request, then proceeded to brief Aster about the White House sessions. "I admit to being a bit unsettled," he confided. "There continues to be substantial discord between Secretary Hurlburt and Paul Vandergrift. You know my concerns about the president's advisor, but now I'm approaching a state of alarm, frankly."

"Care to tell me about those concerns, Admiral?"

"Yes I will. I'd appreciate your counsel, Howard, because whatever we decide here will have a direct impact on my mission to China."

Lee stood, and started pacing. "America is in serious danger of losing its strategic advantages in several arenas. In fact, I'd go so far as to say that there are *no* National Security Council initiatives that capitalize on our remaining advantages. We are in complete reaction mode, stumbling from one crisis to another, like a dazed prize-fighter trying to stagger through the late rounds. There's no *strategic* thinking!"

"I concur." Aster nodded soberly. "And I appreciate your insights, Admiral. It's essential that I know how to read what's going on in Washington, before you head for China."

"The Washington environment is no longer capable of fostering healthy debate, and the lack of trust is palpable," Lee declared. "Howard, the president wants a direct link to us, not through Vandergrift. And this you should know: I'll *have* direct access to the president throughout the China mission."

"What's the going-in premise for your mission?"

"I suggested to President Boyer that we consider a few . . . mmmm, let's say 'nontraditional' ways to improve China's situational awareness. Yes, the Chinese have a limited capability, now, but I'm not sure it's enough to soothe their concerns, especially those of the hard-liners." Lee quickly outlined his *Shoes of the Fisherman* concept, taking note of the general's growing discomfort. Aster abruptly jumped up and stepped to his office's wall-mounted battle screen. He used a remote controller to call up the current operational picture in northeast Asia.

"You know, I respect you a great deal, Admiral. But I tell you; you're treading on mighty thin ice with this old fighter pilot! With all that's at stake regarding China, *what* in the hell prompted you to adopt this way-out strategy?" Aster sputtered, red-faced.

"Because it may be the silver bullet that could stop this space war, and turn around these cascading geopolitical messes."

"Okay, but hear *my* side of this," Aster blustered. "I'm a hardcore military officer, a warrior, just as you were. And a warrior thinks first about his people. Those battle groups and air bases and depots strung across the western Pacific will be nothing but speed bumps against an all-out attack by the Chinese. Hell, and that's not counting pilin' on by North Korea, or whoever else might decide to jump into the fight!

"Your game plan goes against my grain, big time! Every fiber of my warrior's body says 'No'! Are you really committed to this strategy?" the STRATCOM commander asked, incredulous.

"Well, I want to explore it, through gaming . . ."

"Let me finish, Admiral," Aster said, holding a palm up. "Of course, we gotta prepare for critical moves by the Chinese and the

Koreans. Yet, you're telling me that our preps should be an open book? Makes no sense to me. Right now, I'm prepared to go to the mat with Hurlburt over this one!"

Aster had dug his heels in. Lee was reminded of a scene from the old movie *Fail-Safe*, when the NORAD colonel refused to divulge the position of American bombers to Soviet missile controllers. Even with a world war at stake, and a chance to avoid it, a lone colonel defied the president's orders and held an entire situation room at bay with a standard-issue Colt .45 sidearm.

"Howard, Secretary Hurlburt gave the okay for us to game this concept," Lee said. "And I must say he shares the same concerns that you've just expressed. He's firmly on your side of this issue. But he said, 'Let's at least try it on for size during a wargame.' In fact, I'd appreciate having you as an active participant in the China mini-game. We need that warrior judgment."

"Okay, Admiral. I'll keep an open mind—but I'm gonna challenge you," Aster conceded peevishly. "I have to. For the sake of our troops and the nation's strategic well-being."

Aster's communicator chirped, prompting a quick glance at its screen. "Admiral, we'd better get our tails down to the center. The president just ordered our Guam-based bombers to pre-launch alert status."

13 THE PEOPLE'S REPUBLIC

PEOPLE'S REPUBLIC OF CHINA

A four-vehicle caravan of Lincoln Navigators sped away from Capital International Airport outside Beijing. A steady, hours-long rain had delayed and canceled a number of flights, but Admiral Stanton Lee's VIP Gulfstream G550—a stately U.S. government special-mission aircraft—had received immediate clearance into Chinese airspace and radar vectors directly to Capital International. Powerful people were expecting the U.S. emissary.

Almost sixteen hours of travel between Omaha and Beijing had left Lee fatigued. The sophisticated communications suite on the Gulfstream had performed erratically en route, forcing Lee to rely on his own secure-comm "suitcase," with less-than-great results. The suite of equipment combined laser and RF communications gear into a single system that could access the Pentagon's new "Global Information Grid," a worldwide network tying land, sea, air, and space users together.

Due to North Korea's nuclear detonation, satellites the GIG relied upon were decidedly unreliable. Some were completely inoperable, and others were limping badly. Lee hoped Howard Aster had succeeded in convincing the owners of two new Navy spacecraft that Lee absolutely *must* have access to the birds' comm links and imaging capabilities.

In the SUV's backseat, Lee noted his small convoy being picked up by an escort of six Chinese vehicles as they left the airport's perimeter. The mixture of sedans, SUVs, and two canvas-covered trucks threaded its way through barriers and rerouted roads. Construction was evident all across the airport. Runways were being lengthened, tunnels for underground trains burrowed, expanded terminal facilities snaked here and there, and mammoth hangar facilities to house the world's latest jumbo jets rose into the sky. Concrete trucks lined the roads.

Lee's driver broke the relatively uncomfortable silence. "Been a few years since you were in China, Admiral?"

Lee was in no mood for idle chitchat. "Yes. Two-and-a-half years."

The driver glanced at Lee and tried again. "We have about a forty-five-minute drive to the hotel. Rain's keepin' most of the traffic lighter today, so we'll get in just before lunchtime."

Lee nodded and turned to the woman next to him in the backseat trying to get his attention. At the airport, she'd introduced herself as Traci Holden, the CIA's Chief of Station.

"Mr. Ambassador, you met Ronnie Ling back at the airport. He's the Deputy Political Officer. Our driver here is Dan Sheridan, also part of my team."

A faint smile pursed Lee's lips as he nodded at the driver. "Hi, Dan. One request, though, Ms. Holden. Let's just stick with 'Admiral,' if you would, please."

She smiled, more at ease. "May I cover a couple of things with regard to your schedule and recent events?" she asked, glancing at her open day-planner. Lee nodded.

"Ambassador Benson had planned to host you at a private lunch at the embassy, but was called away on urgent business. He's still trying to arrange your meeting with Chairman Yi. Per your wishes, his assistant also booked you a room at the Grand Hyatt in the central district.

"Admiral, about your visit with Chairman Yi; it seems he's stalling, putting it off, for some reason. We'd scheduled a meeting for you to present credentials this afternoon, but the latest official word is he's unavailable until tomorrow, at the earliest."

"I'm uncomfortable with any delays, Ms. Holden. My understanding is the chairman is just as anxious for this meeting as we are," said Lee.

Holden reached into a pouch and withdrew a small device connected by two sets of wires. A tiny microphone and earphone were attached to each end of the leads. She silently extended one end to Lee, placed a molded earpiece into her right ear, then waited for Lee to do the same.

"Can you hear me okay, sir? You only have to whisper." Her voice was understandable, but faint.

"I read you." The device required adjusting one's voice modulation. "Go on."

"Sir, we think the Chinese are *very* confused regarding our intentions," Traci whispered. "There are rumors flying around Beijing about U.S. involvement in an attack on Iranian leaders, but I can't get solid info on that from Langley. Somebody's got that locked up tight. And, of course, the Israelis' attack on Iran has the Chinese quite upset. Following the deaths of Iran's president and religious leaders, Iran's apparently been unusually quiet. Our defense attaché says the Chinese PLA is asking a lot of questions about Iran and our involvement in these incidents."

Shifting her notes nervously, Traci glanced at Lee, hoping he'd offer clarification. If there *had* been an American operation against Iran, the president's emissary to China would know about it.

Lee remained silent. *Typical spook briefing: give you what they have, but they expect something in return.* Lee didn't take the bait.

Traci pressed ahead. "On top of the Iranian strike—or whatever happened there—is some kind of problem related to China's nanosats. I'm sorry, but we're in the dark about that, as well. I assume you wouldn't be here unless you knew about all of these situations, sir. In any event, you may want to discuss these incidents with Ambassador Benson."

No, I won't be discussing these matters with Benson or anyone else at the embassy. Not yet. Lee realized how unsettling his trip must be for the American ambassador and embassy staff. People stationed in a foreign country trusted their own diplomatic family, and embassy people always grew closer in difficult times. But Lee didn't have the luxury of expending time or emotional energy in the name of diplomatic niceties. He needed results, and fast.

Traci continued with her briefing. "Then there's the carrier battle groups. *Abe Lincoln* was on her way home, transiting the Singapore Strait, when she was diverted. Now she's loitering somewhere in the South China Sea. *Ronald Reagan's* group sortied from Yokosuka and is now playing cat and mouse in the Sea of Japan. We know China's keeping track of both the *Lincoln* and *Reagan* groups by flying daily recon missions. Our F/A-18 fighters are intercepting and shadowing China's recce birds every day. They show up like clockwork."

"Makes sense. In their position, I'd do the same," Lee snapped. "Have to keep an eye on any potential adversary in your own back-yard."

"Concerning your mission, sir . . ." she whispered. Setting her notes aside, she turned to face Lee directly. "It's my preliminary assessment that Dr. Zhang has information critical to your mission. And I'm ready to help in any way I can, sir."

"Continue," Lee ordered, still looking askance at her. She hesitated, then crossed her legs and returned to the notes. "Let's see. Ah, yes. A debate is raging within China's Great Hall, or its back rooms, and General Luan appears to be very influential. The PLA is trying to flex its muscles, during discussions about our degraded-satellite plight and the actions of our carrier battle groups. Aggressive PLA posturing has intensified, since we ratcheted up our military readiness on their eastern flank."

No surprise there. Lee had watched the PLA's steady consolidation of power, since the Tiananmen massacre. Following Hu Jintao's fall from power, the Communist Party Chairman, who also served as the president of the People's Republic, had relinquished his chairmanship of the Central Military Commission. That chairmanship now belonged to the military—and General Luan was the undisputed head of China's leviathan military machine, the People's Liberation Army. He had become the most powerful man in China, deftly establishing near-equal status with Communist Party Chairman Lin Yi.

Throughout the previous decade, the United States had positioned itself, diplomatically and militarily, along China's southern, southwestern, and western flanks. Dramatic shifts in alliances, plus military cooperation programs between the U.S. and Uzbekistan, Tajikistan, and Kyrgyzstan, were followed by a formal treaty-alliance with Pakistan, China's weapons trading partner and benefactor of her nuclear expertise. A warming of relations with Vietnam, an ancient enemy of China, and U.S. military-cooperation programs with Laos and Cambodia all combined to feed China's xenophobic nature. Certain Chinese leaders saw the moves as proof that America was pursuing a so-called "containment" strategy. With U.S. Navy carrier battle groups now moving to the eastern flanks, those views were undoubtedly being reinforced.

Of course, Luan sees nothing but containment, Lee glumly concluded. "What is General Luan's status now?" Lee asked.

"That's the sixty-four-thousand-dollar question, Admiral," muttered Traci, shaking her head. "We flat-out don't know."

This is gonna be tough enough as it is, but without Luan— impossible. Lee removed the earpiece. He needed to think. *Welcome back to China. What do we have here? As far as the PLA is concerned, Yi's an interloper. But from Luan's defense viewpoint, our carrier battle groups are just red herrings. He's wondering what we're really up to.*

And why is Yi hibernating at a time like this? Lee struggled to put the diverse elements into a coherent picture. *China remains a paradox, even to its own power factions. Nothing really changes here, not for thousands of years.*

The SUV was creeping through Beijing's tangled traffic, but Lee could see the Grand Hyatt ahead. He exchanged his earpiece and microphone for a thin folder Holden handed him.

"Ambassador Benson prepared this summary for you," she explained. It was stamped "Confidential." The folder contained a single sheet of text and copies of a PowerPoint briefing. "The ambassador had planned to go over this today. Again, he apologizes for not meeting you."

Lee nodded curtly. He was familiar with the politics of ambassadorships. Benson was distancing himself from Lee's mission, in case it turned to mud. Lee turned back to the folder, scanning the

summary page, then the slide copies. He swept through the government gibberish, registering key points:

*China: Initial stages of bloodless coup—Breakaway faction within
leadership, masquerading as "moderates"—Archconservative power
base emerging within Party circles—NSC Director Paul Vandergrift
dismisses findings of a major shift in China power base and its potential threat to U.S. interests—Bold Chinese space program, fueled
by massive growth of China's economy—Social upheaval of great
concern to Chairman Yi and his confidantes—Riots, food shortages,
protests—ignored or minimized by U.S. "experts"—Communist
Party perceives threats from students with proliferating communication technology (personal computers, Internet, wireless communicators, digital cameras, etc.)—Conclusion: Emerging cabal of
"moderates" intends to return China to a cloistered past.*

Lee closed the folder slowly. *God, it may be worse than I imagined! And Paul Vandergrift is keeping Benson's assessments from
reaching the president!* Holden was watching him closely.

"Ms. Holden, please extend my sincere gratitude to Ambassador
Benson," Lee said solemnly. "His summary is most illuminating.
However, it underscores my assertion that a meeting with Chairman Yi is absolutely essential and time-critical. I *must* see the
chairman as soon as possible!"

STRATCOM HEADQUARTERS & ADMIRAL LEE'S HOTEL

The Army Space Command-configured communication link via
new satellites Zulu Zeller had released in orbit was marginal, but
usable, Aster decided. He waited while Army comm experts connected him to Admiral Lee in Beijing. With the loss of so many
communications satellites, Aster's call was being routed through
the Navy's new Arthur and Lancelot multipurpose spacecraft. They
were only available at certain times of day, due to their elliptical,
high-inclination orbits. The double-duty birds also were providing
critical high-resolution overhead imagery of North Korea, China,
and wide swaths of the Pacific region.

The general's eyes scanned a stark summary displayed on his

computer's screen. He didn't like Lee's game plan, but reluctantly conceded the old sailor was probably correct. Hell, the wargamers had confirmed it, at least in the virtual realm. Still, Lee's *Shoes of the Fisherman* strategy ran counter to every bone in Aster's body.

"Admiral Lee here," a disorted voice answered.

"Admiral, Howard Aster here. I trust your mission is going well." Not hearing a response, he continued. "I still have reservations about your *Shoes* strategy, but we've stitched together a system that'll help you play it out. I've confirmed two new spacecraft Zulu Zeller deployed *are* compatible with the comm gear you're hauling around. And, for specific periods each day, those two carrier task groups will be within view of at least one of those satellites. Either of the birds will provide a solid, dependable comm link between you and the boats—as well as overhead imagery you can access."

"My sincere appreciation, General. Is it safe to assume, then, that our national authorities also have reliable, timely communications with the carrier groups' commanders?" Lee asked pointedly. Aster detected both fatigue and tension in the retired admiral's tone. Then again, maybe it was just distortion related to the poor-quality comm link.

"Roger; that's correct. PACOM's validated Arthur and Lancelot comm links, and they've been talking to the *Reagan* task force commander," the general summarized. "I'm supporting you on this, Admiral, but I damned sure don't like it, and I've told Washington why. I've put backup plans in place, just in case you're reading the Chinese incorrectly. You know I have to do that. National security absolutely dictates I take precautionary measures."

"Of course, Howard. In your position, I would certainly do the same," Lee conceded. "And yet, I cannot afford to be surprised here, especially concerning *our* military posture at any given time. Unless Chairman Yi feels he can trust me, and is assured that what I offer is credible and verifiable, then I will have failed, and we are lost. War would be inevitable. Faced with such an unthinkable, would you be willing to share those contingency plans with me? I must know as much about our posture as the Chinese can determine through their own intelligence."

Aster took a deep breath, hesitating. "This *is* an encrypted, fully secure link, I guess." He waited. There was a discernable delay in the communications path, a throwback to the days of early comsats. Finally, Lee responded.

"As secure as the package I brought with me can be, Howard. You know who provided it." The aluminum-cased, portable comm gear STRATCOM's National Security Agency liaison had given Lee was basically the same system covert CIA agents used in the field.

"Good enough for me, Admiral. Here's the short version: I've had the Navy flush the 'boomers'; they're well clear of the carriers, but close enough. All are nuclear-armed, of course."

"As I would expect, General. I'm familiar with the SIOP." The Strategic Integrated Operational Plan was the nation's highly classified blueprint for nuclear war.

"I know you are. The ground-based ICBMs also are on alert, including a few armed with conventional warheads. I can't tell you the subs' or the missiles' targets, but you can guess. Finally, I have several bombers in the air on around-the-clock airborne alert. Others are at the usual Pacific bases; most of those are on one level of alert or another . . ."

Lee interrupted, "Absolutely mandatory precautions you would have to take, I know. On this end, North Korea's aggressive posturing has certainly caught my Chinese hosts off guard, it appears. There have been some rather heated discussions between Beijing and Pyongyang lately!"

"We're aware of them," Aster said cryptically. Signal intelligence capabilities in the Pacific might be degraded, but they weren't completely debilitated. Most of NSA's spaceborne listening posts were still very much in business. "But I can tell you, 'The Dear One' is taking a very aggressive stance. He's deployed one hell of a lot of troops near the DMZ, and the South Koreans have responded in kind."

Lee's response was garbled. The sat-link was degrading. "Anyway, you now have the comm and overhead-imaging resources required for *Shoes*," Aster said, hoping he was being heard. "My people are hustling to get the logistics of ops centers, transportation, and security guidelines in place, should your hosts take you up on our offer. If it's a go, we should be able to accommodate everybody fairly quickly. Of course, our security wonks are going through the roof. I can't blame 'em. This is pretty much what they've spent their lives guarding *against*, you know."

"I understand, Howard," Lee said. "Maybe this will lead to a more-enlightened, far-sighted situational-awareness policy."

"Maybe so, Admiral. That remains to be seen. With people like our esteemed National Security Advisor calling those shots, though, I have serious doubts. But we'll worry about that later. Let's hope your strategy avoids war *right now*—and that the Chinese don't out-fox us and simply use this to blindside the U.S.," Aster warned.

"Concerns noted, Howard. I believe I am dealing with honorable people here—people who want peace, not war, with the United States," Lee stressed. He saw no reason to share his worries about China's palace intrigue and power-struggle machinations at the moment. Aster had enough to worry about.

The general let a protracted silence hang in the air, then asked, "Admiral, are you . . . uh . . . prepared to . . . uh . . . Ah hell! Do you have the means to get your butt out of there, if this plan turns to crap? I do *not* want you on the receiving end of what I may have to send your way. You copy?" the general demanded, his voice a little too husky.

"Rest assured, Howard," Lee said, surprised at the retort. "My embassy escort has made appropriate arrangements, I'm told. How-ever, I'm confident they won't be necessary."

The two men wished each other well and broke the connection. Aster stared from his large office window, unseeing. He couldn't shake an intense, nagging concern that the hair-trigger global situa-tion could still spin out of control and turn into a shooting war. And that could escalate to a nuclear exchange the Earth and its people could *not* tolerate.

He and Stanton Lee were peering into the abyss of World War III, because potential adversaries were operating from positions of un-knowns and suspicions attributable to a lack of knowledge about the others' intentions. As leader of the U.S.' primary global warfight-ing command, Aster desperately wanted to back away from the edge of that bottomless, black pit.

'14 RACING THE CLOCK

STRATCOM COMMAND POST

Dave Forester looked the part of a three-star Army commander, dressed in creased, freshly starched battle fatigues, Aster noted as he entered the command post. The large room's lights were dimmed, enabling the display of complex graphics and tabulated information on large computer screens.

Today, as clouds of imminent war swirled across the globe, dozens of dedicated men and women in battle dress uniforms and flight suits hunkered over computers and moved with purpose throughout the expansive facility. From this room, STRATCOM literally could run a war anywhere on Earth.

Aster stepped into the battle cab and greeted his director of operations. Forester quickly summarized the status of events across the globe, as well as the deployment, alert levels, and overall readiness of U.S. and coalition forces. Aster absorbed the information, asking for clarification now and then.

"Copy all, Dave," he finally nodded. "Exactly where do we stand with the Chinese, from both political and military perspectives?" He pointed at a large screen dedicated to the Pacific Command region.

"In particular, with all the Chinese saber rattling, what sort of missile-defense capabilities do those carrier groups have?"

Forester pointed to a computer screen inside the battle cab. "We now have two locked-and-loaded missile-defense cruisers with those groups, but no Airborne Laser, as we'd hoped. The ABL's still on the ground in California," he said. "But if the North Koreans or Chinese fire a missile, there's a damned good chance we'll shoot it down. The ships' Standard ARM air-defense crews are on twenty-four-seven alert. They just can't hit a missile in its boost phase, like the ABL's laser can."

Aster nodded. "I've ordered our one and only ABL to the Far East as soon as it's back in the air. We may need it—and soon. Look, Dave, I need your take on the rest of this rat's nest," Aster said, waving a hand at the center's primary big-screen. "Anything new on North Korea?"

"The bastards are capable of launching another one, sir. If they're crazy enough to push the button again, they can put a nuke any-where inside that big oval," Forester said, pointing to another work-station screen. Aster whistled softly. A thin-lined red circle covered a huge chunk of the Pacific, encompassing Alaska, the northwest-ern United States, Hawaii, Japan, South Korea, and a sizable swath of China.

"It's a safe bet that the North Koreans won't chuck a three-pointer at China, their only source of food, I'd say. But everybody else in that ring's at risk, sir," Forester said grimly.

"Any more good news?" Aster asked sarcastically.

"Well, multiple intel reports also point to possible North Korean operations against the South, even Japan. Seems our Dear Leader is damn scared, perceiving that his back's up against the wall, so he's shuffling a lot of troops around. Iran assured him a high-altitude nuke would paralyze most of the West, especially us. I suspect Kim knows better by now," Forester half-grinned.

"Yeah, it crippled our space infrastructure, but America didn't roll over and die." The Army general handed Aster a printout, adding, "And, if we buy this CIA analysis, the Israelis have obliter-ated Iran's nuclear capabilities and decapitated its leadership. I doubt if anybody's answering the phone in Tehran's presidential palace these days. Kim Jong Il probably feels a might lonely—and damned vulnerable."

"And we know his Latin buddy, Chavez, is in no position to help the North," Aster added. "Money, fuel, food, and intel from Chavez have completely dried up. Nada from Iran either. These *axis-of-evil* guys build alliances on bluster, but they're all houses of cards."

"As we briefed last night, sir, our primary concern is for the two carrier groups out there in the pond," Forester continued, pointing again. "Those two forces have Kim stirred up. He believes both carrier groups are off *his* coast, preparing to launch overwhelming air strikes. He thinks we can't see shit, either, because he degraded our space capabilities severely. That's what some crafty Iranian agent kept preaching to a North Korean general. Then the phone went dead in Iran."

"And if the North *does* launch?" Aster asked. He'd seen the wargame Red team's latest projections, and didn't like them.

"Pretty much what's been flowing from the game, sir," Forester said. "Our land-based interceptors *might* be able to protect Alaska and the West Coast, but probably not Hawaii. The two carrier groups only have limited organic missile defense capabilities. They can handle sea-skimmers and tactical stuff, but not a strategic missile with a nuke. The carriers are probably dead meat, *if* they're targeted by a North Korean weapon."

"And there's nothing more we can do?" Aster asked, knowing the answer. He'd already ordered, through his Air Force component commanders, several B-2 bombers, a handful of B-1s, and a whole squadron of F-22 Raptors to full alert. None carried nuclear weapons, but all were loaded with conventional armament, some with deep-Earth-penetrating bombs, and ready to launch from Pacific bases on a moment's notice. If that long-range Taepodong left the pad, a hell of a lot of American firepower would be airborne in seconds, most headed for targets in North Korea.

Forester directed the commander's attention to a third wall-size screen. "Sir, as you know, the Chinese are mighty active, too. Intel doesn't know what's going on, but a bunch of people and hardware

are being prepped for something—from air force units to ground-based lasers and submarines."

"Understand. Make sure that type of info is fired to Admiral Lee, as soon as it pops up."

"Lee's still in China?"

"Yep. Sounds like he was hitting speed bumps for a while, but he's finally making some headway. 'Course, Paul Vandergrift is making matters tough for the admiral, but we're staying out of *that* cat-fight," Aster stressed. "Don't be feeding anything up to ol' Van unless I say so. He's *not* in our need-to-know loop, at least for now."

"Loud and clear, sir. I've got enough to worry about . . ."

Aster's handheld-alert was sounding off. The general whipped the unit from its holster, glanced at the screen, saw the security-code text display on his screen, and shook his head. "Washington, finally. I'll check back later, Dave. Stay on top of the Korean and Chinese stuff. That's where the fuses are burning fastest," he added, headed for the door with long strides. He turned to Forester with a final order. "And if things turn to shit in Asia, I want the admiral out of there. *Fast!*"

'15 GATHERING STORM

GRAND HYATT HOTEL/BEIJING, CHINA

The Hyatt's Fountain Lounge hosted a mosaic of power brokers from East and West. They met amid an air of congeniality, but their motives were grounded in money and influence. The campaigns of twenty-first-century deal-making were waged on a global battlefield.

While money was once the universal scorecard, Chinese deal makers also sought prestige and respect for their reemerging nation. In their eyes, China was the new and rightful superpower of a world awash in fanatical death-spirals of religious wars and the decaying financial corpses of colonial powers. Western deal makers sipped tea or bourbon with their Chinese counterparts and saw the latter's Western suits, corporate symbols, and cash, but failed to detect the accompanying hunger and quiet drive for global status.

From his position in the lounge, Stanton Lee watched the dynamics at several nearby tables and sitting areas. He thought about the Chinese businessman who knows the temporary, even fleeting, symbols of Western naiveté. *What about the "moderates" cabal that Benson had described? What kind of deal makers are they?*

Do they tolerate western entertainment and fashion influences as a way to distract their own people, while they quietly accumulate the West's richest technologies and knowledge?

Chinese deal making was all about honoring a 3,000-year legacy. And any insider cabal was merely a self-appointed keeper-of-the-keys that would ensure China's supremacy for the next 3,000 years. These were deal makers of history, not just next quarter's financial results.

"Admiral Lee? Stanton Lee?" a small man asked apologetically. A diminutive woman, carrying a leather briefcase, hovered at his side. The case seemed much too large for the slight woman.

The American stood and greeted the couple, registering their seeming embarrassment or, perhaps, anxiety. The man's suit was too large for a spare frame, and its wrinkled appearance suggested long days of wear. His angular face was pleasant, yet lined with fatigue and a deep sadness. An apathy born of weariness radiated from drooping eyes.

The young woman kept shifting her oversized burden. Stylish, jet-black hair swept across the right side of her face, while a long shock remained fixed behind her left ear, revealing an expensive pearl earring. Her engaging smile was infectious.

"Yes, I'm Stanton Lee, and it is my pleasure to meet . . . ?" queried Lee with sincere politeness. He already knew the man from photographs. Without a word, Lee relieved the woman of her briefcase, gently placing it on the floor next to his chair.

Extending a hand hesitantly, the man responded. "I am Doctor Lin Zhang. This is my daughter, Mi Soo."

Lee gestured them to seats. Zhang sat on an overstuffed sofa that appeared to swallow him. Mi Soo took a straight-backed chair between her father and Lee, crossing her legs in an alluring manner. She shot Lee a smile that abandoned shy girlishness and plainly said, "flirt."

No matter the cultural differences, Lee said to himself. *A smile always sends the same message.* Returning it and chuckling to himself, Lee flashed a mental rebuff. *I'm older than your father, little lady!*

"Such a public place, Admiral. When your embassy people told me where we were to meet—I wondered as to your intent," said Zhang, looking around nervously.

"Doctor Zhang, it wouldn't matter where we met, assuming someone is tracking you. The more clandestine we were, the worse the suspicions. We have nothing to hide. My mission is open knowledge to all interested parties. Besides, this gives our people a chance to see who is interested in you," Lee explained, offering both logic and assurance.

"I fear my father is in danger," Mi Soo said quietly. Lee could barely hear her, again cursing the screaming jet engines that had gradually stolen his hearing over the years. "There are people here who do not trust your country, or your mission. This meeting may serve to confirm their suspicions. You are most-respected by many people in power, but power is shifting like a swollen river that cuts new paths to the sea."

"There are detailed plans concerning America," Zhang explained. "Plans prepared over the last twenty years, maybe. These plans are continually revised, as circumstances change, and they relate to a broad range of contingencies. As you say in your world, the 'what-ifs.'"

"And you may quite naturally assume we do the same with regard to your country," quipped Lee. Years of dealing with the Chinese had honed his own sense of balance. Sometimes you were direct. At other times, obscure.

"Yes, quite so," replied Zhang. "You know my history. Some in your country have referred to me as a 'hard-liner.' But, through others' eyes, I am a 'patriot son of China.' My earlier writings earned me a place in our most-secret planning cycles."

"We are aware of those cycles. I was present at one of the very rare public disclosures, when your spokesperson addressed security planning for the 2008 Olympics. I recall his reference to '*The Way and the Power*,'" said Lee.

Zhang leaned forward, more animated. "Yes, *Tao-te Ching*, our belief in Taoism and its relationship to the martial arts. The mental power that helps one know the way. If you want to know the way, Admiral, you must examine the extremes of the barriers, the opposites of light and dark."

"We do the same in our national-level wargames, Doctor Zhang," countered Lee.

"And that is where we *differ*, Admiral," Mi Soo injected. Her emphasis on "differ" seemed almost personal.

The tiger circles . . .

"You call them wargames, and they are run by your military. We never see them run by your civilian-led government," Mi Soo continued. "Those wargames seem to presuppose only conflict with China. They only examine one extreme.

"My research shows a pattern, a trend, perhaps, which reveals a mind-set that China is your next significant menace. Your traditionalist admirals and generals, backed by the powerful American armament industries, must have a powerful foe. A 'peer competitor' is your term, I believe. But, where is the voice of reason in all these wargames? Your Pentagon is now twenty years, maybe more, into planning for a war with China. The human condition sways to such repetition. It is no longer *will* you fight China? A question. Now America talks only about *when* it will fight China. A given."

Lee's expression revealed a twinge of annoyance. Why dwell on the obvious? They both knew what the Pentagon was thinking. He'd expected more revelation than exhortation from this meeting. Still, the former ambassador drew a calming, deep breath before replying. Long years among the Chinese had bred patience. After all, Mi Soo's summary was as obvious as a tautology.

Zhang glanced at Mi Soo with an ever-so-slight *tsk-tsk* air of mock disapproval. To Lee, he said, "My daughter was also a member of our Center of New Knowledge. She was responsible for studying all findings from your wargames, and for providing our comrades with a view as seen through your eyes. You would know her role as the 'Red team's' perspective. It is this experience that ultimately changed her views and brought enlightenment to me, as well."

"Please explain, Doctor. You speak of your role as if in the past. Has there been a change?" Lee asked gently.

Zhang quickly outlined the Center of New Knowledge and its recent history. "The last two years were trying ones for many of us. We witnessed a change from an atmosphere of discovery to one set on proving a position. Typically, such positions were formulated by a circle of Party members who wish to institute a Maoist version of modern China. They believe a precipitous decline of U.S. influence in Asia is imminent. To them, America's handling of Iraq and Afghanistan was proof that your current president and Congress are weak. China's boldness would fill the vacuum."

Impressed with Zhang's insight, Lee intentionally telegraphed interest through keen attention and gestures, encouraging further revelations.

"Then we found it, the non-nuclear option to crippling America," Zhang continued. "Your space systems. Your satellites and ground stations are such easy prey, as our PLA demonstrated by shooting down our own aging satellite in 2007. Even if deniability failed, we concluded you would not risk a war with China, if we simply removed your space-based military and intelligence capabilities. We see such action as a reasonable self-protection measure, since you've already proved our suspicions about your global-imperialism intents.

"And your policymakers never reached a consensus about space systems. Are they sovereign territory, or not? Is 'spacefaring' like 'seafaring'? Flagged ships are recognized as sovereign soil, but satellites are not? Your intellectuals could never agree on these space issues. To us, such inattention revealed space was America's Achilles' heel."

Zhang paused, expecting a response. Lee only nodded in acknowledgment. "As for Mi Soo and me, we were two of many who argued against the Center's shift from exploration of thought to validating pre-conceived ideas. My daughter argued passionately that we were too casual in our dismissal of American might."

Lee interjected, "So you were asked to leave the Center of New Knowledge, thanks to your heretic views?"

"Well—yes, in a sense. We had also decided that leaving was best for all concerned. Maybe our work could continue at the university, where alternate views might be heard," Zhang explained. "This became increasingly clear to me as I listened to my daughter." The Chinese doctor hesitated, searching.

Mi Soo added, "I was surprised at the quality of debate in your latest wargame series, how vigorously you discussed the opposites, the extremes."

Mi Soo spoke with a passion that belied the Chinese cultural norm of how one spoke to an elder, especially in a first meeting. Lee soon was hearing the familiar words of a passionate convert, the overtones of conviction that comes with a discovery of heartfelt truths, lying dormant, but always there, ready to leap forth.

Mi Soo's dark eyes sparkled. "America is so open! When I first studied your ways, I was disappointed. I saw much confusion in your wargames, so many disharmonies. Even your press reported constant squabbles among your military services and with the outsiders, what you call 'interagency.' I was disappointed, because you were not the coherent, worthy foe I had expected."

Lee wasn't sure whether he should be annoyed or gratified by the woman's honesty. He reached for a glass of water, took a sip to gather his thoughts, then replied, "I'm sorry, but you'll have to explain what you mean by that 'not worthy' statement."

"I am embarrassed to say," she replied, feigning shyness, "for too many years, I heard my seniors speak of the 'shallow American.' No intellect, no curiosity, no substance. I wanted them to be wrong, because I felt they were being deceitful. How could this be? A country that put men on the Moon before we even built a space rocket! Such accomplishments required much intellect and persistence. My own father . . ."

Mi Soo paused, glancing at Zhang and placing a hand on his arm, a gesture of respect. "My own father spoke as so many others did, nourishing the falsehoods of China through contempt for your country. I wanted them to be wrong. I wanted to prove my father wrong. But when I began my research, I saw the same calamity my father saw, a recklessness that accompanied your lack of wisdom."

"What changed? How did you . . . ?" Lee began.

"I read a quote in one of your journals," Mi Soo said. "You Americans have an interesting way of conveying thoughts when your own words are inadequate. You fill your writings with *other* people's ideas. Strange, but I gained clarity through this approach. My father did not believe me at first, but a certain quote set me on a path of truth. My journey along that path continues to this day."

Zhang gave his daughter a look of loving approval, and spoke for her, reaffirming his own conversion. "Louis L'Amour, Admiral. Your cowboy writer. He said, 'To exist is to adapt, and if one could not adapt, one died and made room for those who could.' You see, we Chinese praise ourselves for having a wisdom of the centuries, but that wisdom was born long ago. Such wisdom is missing today. What we dismissed as your simplicity is really your genius. Your country, its people, are always adapting. The debate within your military and your government ensures this adaptability. Even your news media does. We have no such introspection within our leadership today. My daughter showed me the way to such enlightenment."

Lee's handheld communicator vibrated, drawing his attention. Only two people had that number. "I'm terribly sorry. Would you please excuse me? I must take this." He rose and stepped away from Zhang and his daughter.

Answering, Lee was greeted with the familiar refrain of generals that can't, or won't, make their own phone calls. "Admiral Lee, please stand by for General Aster. General, Admiral Lee is on the line. Go ahead."

Aster's voice was hurried. "Admiral, I'll make this quick. Belay what I told you before. Satcom reliability is delta-sierra; it's taken hours to get a call through to you this time. So don't count on *any* satcom, if you really need it. Copy?" He rushed on, not waiting for a reply. "Listen, Vandergrift's trying to torpedo your mission. You probably know that tensions are ratcheting up with the Chinese over space matters— By the way, are you in secure mode?"

Lee responded, "Yes, yes, I am. I can't really speak freely here, though. I've received an update from the Embassy, but I'm not sure they have the complete picture."

"Well, you know Chinese ground-based lasers fired at our space-plane during Speed's XOV mission last year, while he was making a recce run on their nanosats. According to the Naval Research Lab, imagery from one of the new satellites—one of those two dropped into orbit on Zeller's last Blackstar mission—is showing 'a flurry of

renewed activity' around that ground-based laser site in China,"
Aster said. "Look, Vandergrift thinks we've crossed the Rubicon
and that China is poised to launch some kind of strike against our
carriers. Hell, we think the North Koreans are prepping to do the
same! Anyway, Van's screaming that the president needs to take
steps to deter the Chinese from further action."

"What kind of steps?" demanded Lee.

"Preemptive military action against that ground-based laser site,
a couple of nuke-missile sites, command and control centers, sub
pens, and Aerospace City, China's space center-of-gravity. Admiral,
the President is seriously weighing these options as we speak. Hell,
B-2's could be launched from Diego Garcia at any moment!" Aster
spat, agitated.

Lee listened intently as the general related details of STRAT-
COM's planning. If directives were issued by the White House,
Aster was prepared to give "weapons free" orders to the two carrier
battle groups in the Northwestern Pacific, clearing them to take on
all Chinese submarines that sortied east of the outer island chain.
All Chinese military aircraft within 500 miles of the carrier groups
would be deemed hostile. To protect America's increasingly scarce
reconnaissance satellites, all of China's ground-based laser sites
had been declared threats and were targeted for destruction in a
first-wave strike.

Aster paused to catch his breath, then added, contempt lacing his
words, "That SOB Vandergrift recommended launching a batch of
conventional warheads on ICBMs and sub-fired Tomahawk missiles to complement B-2 strikes! 'Saturation and widely diverse delivery methods would confound Chinese air defenses,' he's saying!
This guy's trying to start a war on his own!

"None of this came out of our wargaming, either. In fact, once
SecDef told me what Van was proposing, we 'gamed it—hastily, I
admit—but the outcomes were disastrous!"

The STRATCOM chief proceeded to outline China's responses,
as predicted by wargamers: Six nuclear missiles armed with relatively low-yield warheads would target U.S. military bases in
Guam and Okinawa. Hundreds of conventional missiles would

strike Taiwan, targeting military and political leaders. The *Lincoln* and *Reagan* battle groups would be hit with rolling cruise-missile attacks from submarines, aircraft, and land-based launch sites. The Chinese would employ every measure available to sink all ships in both U.S. Navy battle groups.

Wargame outcomes suggested that, in the midst of this action, China's leaders would present President Boyer with a démarche, backed by threats of further escalation. Any retaliation by the U.S. against Chinese sovereign territory would be met with nuclear strikes against Hawaii; Bremerton, Washington; San Francisco; Long Beach; San Diego, and Las Vegas. The démarche would include public disclosure of Chinese intentions before the United Nations General Assembly, where China also would denounce the U.S. as "the original aggressor" that prompted a response. Specifically, China would reference "an unprovoked attack on Chinese satellites, followed by preemptive strikes by B-2s, Tomahawks, and conventional ICBMs," Aster concluded.

Lee felt time and space constricting, trying to smother him. He needed that damned meeting with Chairman Yi! And he needed to speak to the President, now, before any B-2 bombers were launched.

He glanced over his shoulder at Zhang and the man's daughter, who were engrossed in animated conversation. Whatever avenue to peace Lee and they might have opened within the next few hours could be destroyed in minutes, the path covered in blood and nuclear debris. At immediate risk were two aircraft carrier groups, six American cities, and probably Beijing, Shanghai, and most of North Korea. All could be smoking, radioactive ruins, if nobody stopped this madness.

The Russians would go berserk, of course, and Prime Minister Dmitry Zubkov, Vladimir Putin's puppet, would launch those rust-bucket Russian nuclear subs, given the slightest excuse. Russia's command and control systems were decrepit, at best, so who could predict what those subs' skippers might do? Lee had to break through the wall of falsehoods Vandergrift had erected, effectively insulating the American president from the real world, a world on the verge of disintegration.

WHITE HOUSE SITUATION ROOM

President Pierce Boyer swept into the White House Situation Room and took his place at the head of its long conference table. Only four others were clustered at his end, leaving the room virtually empty. Their grim expressions said this was no routine gathering.

"Herb, I know you've already run this by most of us here, but give us the latest. I want everybody working from the same baseline," the president ordered.

National Intelligence Director Herb Stollach tapped a few keys on a slim laptop computer, bringing up an image on the room's wall display. "Not much has changed since I spoke to each of you. This North Korean Taepodong appears to be ready for launch, and we're ninety-nine-percent sure it's armed with a nuclear warhead. Target's unknown, but we're betting those two carrier groups steaming toward Southeast Asia are prime candidates. The Taepodong is notoriously inaccurate, but, with a nuclear airburst, close is good enough."

"T.J., what defensive measures are being taken by those carriers?" Boyer interrupted.

Hurlburt, hunched over folded hands, delivered the unvarnished news. "All hands are at general quarters on every ship. Our subs are down deep, but might still be affected by an air burst. We've flushed the nuclear boomers, too. They're well away, in case we have to respond in kind. Most of our aircraft are below-deck on the two carriers, although we gotta keep a few topside, on alert. Of course, several fighters, tankers, and the Hawkeyes are airborne, per standard procedure."

"What about the sailors and marines? Anything more we can do to protect them?" Boyer pressed.

Hurlburt shook his head. "Nothing more than we've already done. They're wearing protective gear, but everybody has to be at his or her battle station, and that means some are exposed. Nonessential personnel are where they're supposed to be, in the ships' interiors, but they all have damage-control duties. They'll be outside in a heartbeat, if their boat takes a hit. No, Mr. President, everybody's pretty much in nuclear-KYAG mode and they know that's as good as it gets."

Boyer looked puzzled, as did all but Herb Stollach, who was pointedly studying a fingernail. "I'll bite, T.J. What the hell is KYAG?" Boyer growled.

"Sir, GIs are taught that, in the event of a nuclear attack, they're to bend over, put their heads between their knees, and 'kiss your ass good-bye.' That's KYAG," Hurlburt deadpanned.

"Good Lord! You're sick, Hurlburt!" fumed Paul Vandergrift, slamming a palm on the table.

Boyer cut him off. "Thanks for enlightening us ignorant civilians, T.J.," the president retorted sarcastically, "but at a time like this, sophomoric comments are *not* helpful. Korean crazies are poised to go nuclear *again*, Iran's crop of fanatic mullahs may or may not be muzzled, and the Chinese are up to God-knows-what. Our space situational awareness is improving, but we're still staggering. Does that about sum up this mess we're in?"

Stollach and Hurlburt traded glances, before Hurlburt answered. "That's right, sir. In critical areas of the world, we have fairly good SA, at least for the moment. It's spotty, we've lost some commercial satellite communications we'd normally be tapping, and the entire space infrastructure's mighty precarious. General Buzz Sawyer's been able to get a few interim capabilities into orbit, using new space-tourism commercial-launch outfits. And, as you know, Zulu Zeller put up two multipurpose birds on his tragic mission. Things could be a lot better, but the space network's in fair shape, considering."

"Mr. President, I don't know where T.J.'s getting *his* rosy glasses these days, but I sure as hell can't find any," Vandergrift exploded. "I assume we're reading the same situation reports and getting the same intel. From my perspective, we're in an extremely vulnerable state, and our ability to monitor the global situation is terribly deficient. Yes, North Korea is an immediate threat, but China's the one that really warrants our attention and top-priority, full-spectrum monitoring. We cannot trust those Communist bastards! They say one thing and do something else, smiling and bowing the whole time! I think they're up to something, and it's *not* in our best interests!"

Boyer indulged the outburst, then raised a hand. "I hear you, Paul. But just what do you propose we do about China? Ambassador Lee's over there as we speak, and he's scheduled to meet with Chairman Yi very soon. I think we should see what comes out of Lee's efforts before we consider other courses of action."

"And General Aster's wargamers are working on those other options right now, sir," Hurlburt added, searing Vandergrift with a dark scowl.

"Yeah, yeah." Vandergrift flicked a hand, dismissing both comments. "But we're putting our forces and citizens at considerable risk by giving China time to position forces, then act, while Lee's listening to the chairman's platitudes of peace. Far as I'm concerned, China fired the first shot when they hit Speed Griffin's . . . uhh . . . space 'hot-rod' with their friggin' ground-based lasers! And the Chicoms *had* to be behind that North Korean nuke blast! They're probably urging Kim to fire a Taepodong at our carrier groups!

"Do we stand by and let them get away with *that*?" Vandergrift demanded. "I tell you, we should strike immediately. Hit them where it hurts, hit them hard, and hit them now, or we'll soon be apologizing for thousands of American casualties. We simply cannot afford to wait for China to play its cards first! The stakes are far too high!"

That triggered outbursts from Hurlburt, Stollach, and Gil Vega, the president's chief of staff, each trying to outshout the others. Boyer shook his head and held both palms in the air. The turmoil finally quieted. Boyer looked around slowly before speaking.

"Gentlemen, and I use the term *very* loosely, given what I've just witnessed, Paul makes some good points. I'm not prepared to launch a preemptive strike against China, but that's not off the table. T.J., make sure Aster's people take a hard second look at the ramifications of such a strike, as well as the downsides of our taking a hit from the Chinese first, *then* responding. And I hope they're looking at *all* areas of concern. Not just military ones. Do I make myself clear?" He stared at Hurlburt, who nodded.

Hurlburt barely heard the subsequent back-and-forth chatter, mostly "administrivia" by his standards. He needed to get a heads-up message to Stanton Lee, and fast. That damned Vandergrift was up to something. . . .

LEE'S HOTEL LOBBY/BEIJING, CHINA

Lee keyed in a secure-line White House number and checked the hotel lobby, while waiting. Zhang and Mi Soo were standing, their heads close. The woman was agitated, it seemed. Lee stepped outside and stood in the hotel's circular drive. Nothing. *Damn! Is that satellite out of range already?*

He disconnected, then dialed the American embassy and asked that a vehicle be sent to pick him up immediately. Somehow he had to get to Chairman Yi. He broke the connection and again selected the White House code. Finally, the President's chief of staff answered. "Vega here. What can I do for you, Admiral?"

"You know why I'm calling, Mr. Vega. Please, I must speak to the President. Something very important . . ."

Vega interrupted, speaking so softly that Lee strained to hear him. "The President is weighing several options proposed by Vandergrift, but opposed by Hurlburt. The problem is time and knowledge."

"Knowledge?"

"Yes, knowledge. Van has convinced the President that *our* lack of situational awareness, due to all these satellite outages, has put America at great risk, and we must act first. Next, those Chinese snooper-nanosats have really unnerved some of the president's advisors. And, I must tell you, Admiral, China's urging North Korea to detonate that nuclear weapon at high altitude didn't help matters *at all.*"

Lee protested, "But the Chinese aren't behind the North's detonating that nuke! Yi did *not* condone North Korea's actions. He . . ."

"The President is available now," Vega interjected formally, cutting Lee off.

"Boyer here. How's it going, Admiral?" asked the president, unexpectedly chipper.

"Greetings, Mister President. I'm meeting with Doctor Zhang at this very moment. He's definitely concerned about aggressive posturing by some Chinese leaders, and he's worried about how they may perceive our intentions. My meeting with the Chairman was set back, but I'm working to reach him soonest. However, that may require a hotline call from you to . . ."

"Listen, Admiral, I'm in a damned swamp here," Boyer snapped. "Yi hasn't seen fit to meet with you or to call *me*. We may have to deal with an attack by the North Koreans. Maybe the Russians, too, before this is over. But at this instant, it's the Chinese I'm worried about. Intel says they're posturing to inflict horrific damage on us, and I won't sit back, waiting to take that first punch."

Lee chose his words carefully, trying to avoid further ire. "I understand, sir. With due respect, may I add that I'm sure Chairman Yi is thinking similarly? We *must* find a way to step away from the brink! Our original premise still has merit. China won't act before Yi meets with me, I'm sure. They will let *us* make the first mistake. Any aggressive action we take at this juncture *will* be a mistake, a serious mistake, sir," Lee stressed.

"I *must* have an opportunity to present myself and speak to him in a way he will understand," he implored. "If we demonstrate goodwill, with tangible checks and balances, we stand a chance of achieving a mutually acceptable outcome and avoiding disastrous consequences. I just need to get Yi to buy into our strategy, backed by your personal assurances."

The only response was protracted silence. Boyer faced awesome responsibilities, and the options available ran counter to his idealism, challenging his core beliefs. He was pulled in confusing directions, Lee realized. Boyer's trusted *civilian* National Security Advisor was pushing for aggressive action, supposedly in the name of deterrence. Yet Lee, an equally trusted emissary, an ex-*military* officer, was urging restraint, favoring diplomacy.

"What if I only order strikes on China's ground-based laser sites? Will Yi take that as a stern warning? See them as a firm message of our resolve?" Boyer offered hopefully.

"Not a chance," Lee said without hesitation. "We will have forced Yi into escalating tit for tat, Mr. President. Besides, Vietnam proved the futility of the White House sending 'messages' to an adversary through selective bombing. In this case, such a 'message' would only bring tremendous pressure on Chairman Yi to take that next step. And he wouldn't be able to resist the hard-liners for long.

"Yi didn't become chairman by being rash. But to survive politically, he'll have no choice but go down the road of escalation. Sir, I'm sure you've been briefed about STRATCOM's wargame findings concerning such escalation." Lee slowed, letting his words register. "We lose. Politically at first, then possibly militarily, if we push the path-of-no-return button."

"Yes, I've been briefed about the 'game outcomes," the president acknowledged. "But I can't depend on some academic exercise! Not now! Real-world decisions must be made to protect the American people. Understand?"

Lee's spirits sank. *'Can't depend on some academic exercise'* . . . Desperate, he shot back, "Nor can we afford to make bad or rash decisions, sir! Those wargame consequences were derived by very smart people immersed in a highly sophisticated crucible of knowledge and understanding. Their methodology has taught us how to adapt to critical situations, and the Chinese appreciate that. I guarantee they do. I implore you to take note of the 'gamers' findings, and let me complete the mission you assigned to me."

Still on the line, the president's chief of staff chimed in. "Pardon me, Admiral, but until Chairman Yi communicates with us, we simply don't know China's motives. We're getting intel assessments indicating the PLA was behind the North Korean decision to detonate that nuke, as well."

Undeterred, Lee shot back, "That's contrary to very credible CIA and NSA intel showing *Iran* was behind the North's decision. Mr. President, I must . . ."

"Six hours, Admiral," Boyer interrupted brusquely. "I'll give you six hours to get Yi on the hotline. Go ahead and present our more-than-generous offer to avoid war, based on your plan. But make this perfectly clear to Yi: My finger's definitely on that go-to-war button. In the meantime, I will advise Ambassador Benson to take necessary precautionary measures to ensure embassy and staff safety. You'll need to do the same. Good-bye, Admiral. And good luck."

Shaken, Lee snapped the communicator's cover closed, unsure whether he still had the President's full confidence and support. It may not matter, though. They were now into those delicate moments, where either leaders' decisions would trigger monumental, Earth-shaking upheavals, or secret liaisons and the actions of ordinary men would avoid cataclysm.

Six hours! Would it be enough?

Returning to the hotel lobby, Lee spotted Mi Soo, speaking into her handheld near the Fountain Lounge doorway. She failed to notice him. Zhang was nowhere to be seen.

Now what? Lee thought, searching for the slight man, struck by a growing sense of dread. *Where could Zhang have gone so quickly? Can the daughter be trusted? Are these two delaying me on purpose for . . . for what?*

Mi Soo angled across to intercept Lee. She drew close to him, then whispered, "Chairman Yi will see you in one hour. Unless you object, the Chairman has asked that I escort you."

The young woman's proximity caught Lee off guard. A faint scent of expensive perfume floated between them. How could one Mi Soo phone call have accelerated his meeting with China's top leader? *Am I being set up . . . ?*

"That's great news. And—I'm more than delighted to have your company, Mi Soo—as an escort, of course," Lee stammered.

"My father asked that I convey his regrets. He was called to make some required updates, and only he holds the information they need," she explained vaguely.

"Updates? Information for whom? Who are you talking about?" he asked sharply, following the woman outside the hotel. He hesitated, thinking quickly. "Mi Soo, please allow me to freshen up a bit. I'll just run up to my room and meet you back here in five minutes."

The woman stepped close again and whispered emphatically, "There is no time! Nu and his 'moderates' are even now taking provocative actions that will give your president no choice but to attack China!" Her dark eyes held his, unwavering, yet worried. Was she trying to explain her father's rapid departure? Or was it something else?

Skeptical of her motives, Lee rejected the internal alarms her demand had set off. "I need to place a call to the States. I have to ask you to wait here."

"There is *no* time, Admiral! We must get to the Great Hall as soon as possible!" she stage-whispered, grasping his arm.

He stared into her dark eyes, which never wavered. Gut-level intuition told him to trust this slip of a woman, although his brain screamed otherwise. "All right. We'll go. An embassy car should be here momentarily."

Mi Soo's anxiety seemed to ease. She glanced around his shoulder, then muttered, "Chairman Yi is seeking support at this moment. But you must be well-prepared, before you meet him. Once we enter the Great Hall, our fates are linked to forces we do not control."

'16 TREASON

IN ORBIT, OVER EASTERN CHINA

Seated in a nondescript, windowless mission-control center at the Naval Research Laboratory near Washington, D.C., a junior engineer confirmed the lab's new Lancelot minisatellite was on track, west of Beijing, China. He double-checked a well-thumbed flight-plan booklet, verifying the multimission satellite's compact suite of visible-light, infrared, and miniature synthetic-aperture radar was automatically scheduled to acquire surveillance data on this orbit.

Lancelot's integrated electro-optical system was aimed at a prime target highlighted on that flight-plan list, unblinking optics prepared to sweep across a multicolored expanse of China. Light-sensitive detectors noted the Sun's angle, and automatically adjusted the optics train to ensure only a measured quantity of visible light and infrared photons fell on solid-state sensor arrays buried in the spacecraft's tightly packed interior.

The NRL operator had just turned away, when a loud audio alarm startled him. He snapped back to the oversized workstation screen marked Lancelot and quickly scanned a stream of red-tinged data. "Oh, shit!" he breathed, grabbing a nearby portable phone and punching a speed-dial key.

"Warner," a bored voice answered.

"Jim! Lancelot is being painted by a tracking radar! Ohhhh *shit*! Now it's gettin' blasted by laser energy!"

"What the . . . ! Where's the bird?"

"Uhh . . .'bout twenty klicks west of Beijing. *Holy* . . . ! The bird's being hammered by laser shots!"

"Did the auto-protection activate? Make sure the covers are . . . !"

"Yeah! Looks like the threat sensors did their thing. . . . Uhh . . . No sign of damage to imaging systems; onboard radar's still transmitting and receiving. . . ."

"Get the whole platform into 'safe' mode! Don't wait for the auto-system to do it! Send a 'safe' command, *now*! I'm on my way down!"

The operator was sweating. *Holy . . . ! Our baby's under fire!* his brain screamed. He ripped a plastic-covered sheet from a clip on the cubicle wall, ran his finger down a long list of arcane commands, then started typing. He carefully entered a series of short commands that he had believed would never be sent to the twin satellites, Arthur and Lancelot, once they were in orbit. He and his colleagues had trained for this kind of low-probability event, of course, but nobody believed a satellite would really be attacked. But Lancelot had been.

In the dark cold of space, Lancelot's onboard threat sensors had performed flawlessly, automatically safing the vehicle in a split second. Initially, a radar detector had registered the spacecraft being painted by a ground-based search radar. Seconds later, it noted the telltale high-pulse-repetition frequency of acquisition-and-tracking radar pulses, which triggered the alarm in NRL's mission control center.

Before the first blast of laser energy slammed into the satellite, guided by those tracking-radar signals, Lancelot's auto-safe systems had effectively squeezed optical apertures closed, snapped protective composite covers over delicate lenses, and rotated both its solar cell-covered wings to an edge-on orientation. In essence, the craft had curled itself into a ball of self-protection, braced for whatever followed.

Tiny laser sensors scattered across the satellite's Earth-side surfaces were subjected to intense shocks of short-duration laser pulses powerful enough to destroy roughly half their number. Carbon-composite structural panels angled away from the Earth did their job well, deflecting the bulk of laser energy away, while ablative surfaces dissipated the remaining energy. Surviving laser detectors transferred critical signals to onboard microprocessors, which delivered their analyses to NRL operators in seconds: A Chinese ground-based laser near Beijing had attacked Lancelot, trying to destroy the vehicle. But state-of-the-art military spacecraft design had parried the brief attack successfully. Onboard systems remained functional.

Undeterred, the spacecraft climbed higher and higher, following an elliptical orbit that, for several more hours, would keep the vehicle's transponders in position to relay national security communications to and from the Northern Pacific and Asia. The inanimate space traveler was unaware of the lifesaving role it was about to play in the drama unfolding far below.

GRAND HYATT HOTEL/BEIJING

Spotting a U.S. embassy Lincoln Navigator turning into the hotel's curved drive, Mi Soo anxiously tugged on Lee's sleeve, drawing his attention. "You must know . . . my father is working with General Luan," she began, "but it's not what you think, Admiral. General Luan is a Chinese patriot, like my father. We are passionate about our country, our people, and our destiny. . . ."

The Navigator stopped twenty feet from the hotel doorway, and Traci Holden, Beijing's CIA station chief, stepped from the front passenger seat. She opened a rear door and waited. Taking the Chinese woman's elbow, Lee urged her toward the vehicle. "Mi Soo, please. We have no time to discuss laments . . ."

She cut him off. "Your meeting with Chairman Yi can seal the fate of many good men and women. Moreover, there is one in your country who *must* be confronted."

Lee's eyebrows rose in question, but he kept walking.

"Paul Vandergrift, Admiral," she hissed. "He is coordinating with Nu, but Vandergrift doesn't know he is being used."

The retired admiral gave no outward indication that he already knew about the U.S. National Security Advisor's backdoor dealings. Aster had told him, before Lee had departed for China. But Mi Soo's statement triggered a flash of insight. *Of course! Zhang and Luan. Strange combination. But Vandergrift and Nu! Dangerous . . .*

Lee knew enough about China's arcane power struggles to not assume anything. The party elite had long ago learned the tradecraft of exchanging people for personal gain, often violently. Even as China's quasi-capitalist economy flourished, Maoists endured and remained influential. They still relied on the heavy-handed methods of Bolsheviks, Nazis, fascists, and jihadists, as well. "The Cause" was always foremost. People were secondary and expendable. Lee could only hope that Zhang and Luan were pursuing an enlightened path, but he had no idea where it might be leading. He *did* know, however, that no good could come from a Vandergrift-Nu nexus.

"Mi Soo Zhang, Traci Holden," Lee clipped as they reached the Navigator. "Traci, Mi Soo is accompanying me to the meeting with Chairman Yi."

"All bets are off, concerning a meeting with the Chairman, sir," Holden interrupted tersely. "Ambassador Benson wants you to return to the embassy immediately. Something big's going down. We're not sure what, but these streets could soon be bad-guy country."

Lee shot Mi Soo a questioning glance. Eyes wide, she nodded assurance. "Take us to the Great Hall, as planned, Ms. Holden," Lee ordered tersely. "I'll release you from any responsibility regarding my safety, and you'll be free to return to the embassy."

Taken aback, Holden eyed Lee and the Chinese woman slipping into the Navigator's second-row seats. Lee touched Traci's shoulder, as the station chief buckled her shoulder belt. "Ms. Holden, the President updated me about the North Korean situation, and indicated that all pertinent info would be available through the embassy. Do you have any back-channel inputs for me?" She turned in time to register Lee's surreptitious wink.

Thanks, Admiral, Traci thought. *Good to know you're still on our side!* She took the subtle reassurance as an opportunity to toss a dismissive, smug glance toward Mi Soo, then handed Lee several papers. "Yes, I do, sir. These came in, just as we were leaving the embassy. Jill Bock, some STRATCOM analyst, asked that we get this summary to you ASAP. She called it a BOYD something or other. It's an update of space assets available for . . . uhh . . . She called it, 'the situational awareness thing.'"

"Sounds like Jill," he chuckled, taking the sheets. "Thanks."

"She also included these China-game updates, but . . ." Holden added, holding a second sheaf of documents aloft. The cryptic statement was left hanging, obviously for Mi Soo's benefit.

Scanning the BOYD summaries, Lee mumbled, "Could you capsulize the game's findings and recommendations? Probably on the last page. . . ."

"Admiral, is *she* cleared for this?" Holden demanded. "I mean, she's . . ."

"She's cleared," Lee snapped glaring at the CIA chief. "I cleared her. Please, the findings and recommendations."

Holden's face reddened as she turned away. Noticing Lee's hotel ballpoint had failed to write, Mi Soo handed the white-haired man a marking pen, accompanied by an engaging smile. He started sketching diagrams in the sheets' margins, converting BOYD's textual analysis of ISR coverage into graphics. He glanced at Holden, questioning her hesitation.

"Okay, here goes," Traci shrugged, annoyed. "Oh, before I get into the findings . . . ," she interrupted herself, pointing at the documents Lee held. "You'll find separate dossiers in there covering Feng Bao Nu, Doctor Zhang, and General Luan. Your BOYD-trick analyst says you should keep 'em bundled."

"Pardon me?" Lee queried, puzzled.

"Bock says they're connected somehow," she shrugged again. "The embassy knows all three of 'em, but we've never put them in the same circle of associates. Frankly, it looks odd to me. I don't see the connection."

Twisting to half-face the rear seat, Traci summarized the wargame outcomes. Lee listened, nodding now and then, but continued his review of satellite-coverage data. *Well, the findings and recommendations from that game's last few moves certainly underscore the value of my mission*, he decided, as Holden wrapped up her comments.

Throughout Traci's rundown, Mi Soo listened closely, apparently familiar with wargaming terminology: intent, findings, issues for debate and consequence, range of options and recommendations for operational objectives, then diplomatic strategies.

The Americans have become very good, she thought admiringly. *Wargaming is now their realm; they have built their* own *center of new knowledge. So enterprising is their culture!*

STRATCOM HEADQUARTERS/COMMANDER'S OFFICE

Returning the secure phone to its cradle on his desk's command-center-like communications panel, General Howard Aster stared at the Nebraska plains and tried to shake off a wave of despair. He couldn't help but wonder whether he'd see a new morning through that office window. He was stunned, struggling to make sense of what he'd just been told. A Chinese ground-based laser site had attacked the Lancelot spacecraft Zulu Zeller had placed in orbit only days earlier. The same damned GBL battery that had zapped Speed Griffin on his last Blackstar flight.

Alone in Beijing, Stanton Lee was depending on Lancelot and its sister satellite, Arthur, to carry out his "shoes of the fisherman" strategy. If either of those spacecraft were knocked out, Lee would be isolated, cut off from STRATCOM and the White House.

Given the tenuous balance between China and the U.S. at the moment, why would the PLA take such provocative action? Aster couldn't pretend to understand the Chinese Army's reasoning, but that didn't change the orders he'd been given.

When informed of China's laser attack on Lancelot, President Pierce Boyer had immediately ordered a "measured retaliatory strike," dismissing strong recommendations to the contrary offered by SecDef Hurlburt. Aster had reluctantly carried out the directives of his civilian leaders, launching several stealth bombers from Pacific bases. The low-observable, long-range B-2 Spirits would target every known ground-based laser complex in China, not just the guilty site near Beijing that had zapped Speed's spaceplane. Somehow, that idiot Vandergrift had convinced the president to launch a strike against those GBLs.

There was absolutely no question as to what those strikes on China would unleash. STRATCOM wargamers had beaten that scenario to pieces. Outcomes were always the same: U.S. retaliation would spark a rapid tit-for-tat military escalation that culminated in a devastating U.S.-China nuclear exchange.

Validating the wargame results, an independent assessment by Major General Dawn Erikson and her NSSI team had presented stern arguments against a strike of *any* kind. A member of NSSI's cadre had recalled discussions with a Chinese space-operations counterpart two years earlier. At the time, the Chinese officer was bemoaning America's dominance of space, saying China was concerned that Pentagon budgetary decisions were underwriting an aggressive buildup of U.S. military space forces. In response, the officer claimed, China had been forced to deploy ground-based lasers as a defensive counter to America's space power.

Dawn Erikson had warned, during the NSSI cadre's outbriefing to Aster, that any attacks on those GBLs would disrupt a delicate balance of power. China would interpret the strikes as a deliberate attempt to neutralize a strategic-defense asset prior to an all-out U.S. attack on China, and would respond accordingly.

SecDef Hurlburt had conveyed those wargame and NSSI conclusions to Boyer, but to no avail. During his last telecon with Aster, Hurlburt had pretty well summed up the situation, as he signed off: "If Van believes this Nu sumbitch can single-handedly stop the PLA from launching a massive counterattack, after we bomb the stew out of their GBLs, he's out of his frickin' mind! This *will* go nuclear, Howard. Count on it!"

EN ROUTE TO GREAT HALL/BEIJING

Holden's report about the improved state of U.S. space, air, and naval assets throughout the Pacific theater raised Lee's spirits markedly. The two Navy carrier battle groups in the Northern Pacific were relatively unscathed by the North Korean nuke's EMP, due to nuclear-hardened equipment. And at least one of the two NRL satellites recently orbited by the ill-fated XOV spaceplane would be in position to carry out his plan—*if* he could get to Chairman Yi within the next couple of hours.

Traci concluded the STRATCOM-forwarded wargame summary, then switched to updated intelligence reports. Glancing at Mi Soo, Holden said, "Back-channel sources confirm South Korean preps for massive air assaults against People's Republic of Korea nuke-missile launch sites and political-military headquarters. The South intends to take 'em out with iron bombs dropped from whatever flies. Also, it looks like the South has more than five divisions of troops in position to cross the thirty-eighth parallel.

"Seems North Korean forces along the DMZ were rendered deaf and blind by their own nuke blast—far worse than the South's were—so the North's troops are considered highly vulnerable," she continued. "From what our analysts can determine, the South only wants to secure the North's tunnels and any artillery that threatens Seoul. Supposedly, they'll stand down at that point."

Lee absorbed the torrent of information, integrating it into a reformulated outline of his negotiating points. *As expected, things always change. New challenges, new opportunities. Need to get China to lean on the North Koreans, while President Boyer convinces the South to stop all military operations. We can't let those damned ROK planes launch! A U.S. priority must be defusing the Korean peninsula, so Yi isn't distracted by that dustup.*

Street approaches to China's Great Hall were nearly impassable, choked with army trucks, armored vehicles, and security personnel. The PLA definitely was ramping up preparations for war. The Navigator's American flag and diplomatic plates were sufficient to ensure passage at most checkpoints. Mi Soo's government identification helped clear the others, a point not lost on Traci Holden.

But their luck eventually ran out. The embassy driver started cursing the increasing number of obstacles that impeded his SUV's progress. Traci finally declared, "To have any hope of making it on time, Admiral, you'd better go on foot from here. It's not far. And don't forget your package."

The driver jumped out, opened the vehicle's rear hatch, and retrieved a metal case containing Lee's miniature command, control, communications, and intelligence center. About the size of an airline carry-on, Lee had asked Holden to keep it at the embassy.

"Good luck, sir," Holden clipped, offering a hand. Lee thanked her, then followed Mi Soo onto the crowded walkway. They made little progress, trying to push through throngs of people, until she recognized a senior PLA officer and asked for an escort to the hall. The officer's rank could only be determined through other soldiers' deference.

Their meaningless egalitarian practice of forgoing rank insignia lives on, thought Lee.

At the Great Hall's official entrance, a knot of civilian and uniformed personnel awaited them. Lee took Mi Soo's hand and whispered, "I have been warned about a split within your government. Some of your former wargaming friends want to fight, and do it now, while the U.S. is supposedly back on its heels. General Aster confirmed the U.S.-China situation is worsening. And I'm aware that Feng Bao Nu is collaborating with Paul Vandergrift to exacerbate this rift. I fear something devastating is . . ."

"Thank you, Admiral. Most kind of you. Yes, I'll be happy to join you for the meeting with our Chairman," Mi Soo covered, loud enough for the approaching Chinese security detail to hear. As she let her hand fall from Lee's touch, she leaned closer and whispered, "It is Nu's cabal; *they* want this fight! But they have tricked America into a first-strike against China, for the world to see. Nu's cabal believes your aggressiveness somehow will give its actions legitimacy, clearing a path for Nu to seize power. I do not understand such crazy thinking, but this *is* a palace coup!" she stressed.

Lee was again taken aback. It was as if Mi Soo had overheard his conversation with Aster. STRATCOM's wargamers had predicted precisely this type of internal power play by a renegade Chinese faction.

WHITE HOUSE/OVAL OFFICE

President Pierce Boyer was a wreck, T. J. Hurlburt thought. True, it was damned late—or very early—and the prez wasn't exactly a night owl. Being dragged from bed two nights in the same week by his SecDef and National Intelligence Director had left the world's most powerful man grouchy and haggard. *The guy isn't handling pressure real well*, Hurlburt concluded silently. *And now this.*

Boyer paced back and forth across the Oval Office, repeatedly running a hand through sleep-mussed hair. "T.J., I just can't believe Paul Vandergrift would do something like this. I've known the guy for—well, a *long* time! This isn't like him! I really have to question whether those digital recordings are bogus, faked to get at Paul for some reason. Couldn't this be something your wild-assed generals cooked up? Some kind of brazen Pentagon coup against my administration?"

Ahh, shit! Now we gotta deal with palace-guard paranoia? Hurlburt fought for control, but an almost-explosion was stifled by a firm grip on his forearm. Herb Stollach stepped in before T.J. could say something he'd surely regret.

"Mr. President, I personally listened to the raw recordings of Paul's and Feng Bao Nu's telephone conversations intercepted by NSA. Without the gritty details of where and how we snagged them, I can absolutely assure you they *are* valid. Your national security advisor is working a backdoor deal with a Chinese intelligence agent. And the most benign thing I can say about their deal is that it runs counter to your policies and America's best interests. No question about it."

"Come *on*, Herb!" Hulbert objected. "You know damn well that constitutes . . ."

Stollach's hand, still on Hulbert's forearm, squeezed tightly, signaling: *Shut up! Let the president come to that conclusion on his own!*

"But why, for God's sake? To what purpose? Paul's no damned traitor! He's *got* to have a good reason for doing this! I know it! I feel it in my bones!" Boyer was almost whining, wounded deeply, betrayed by one of his most trusted advisors. Compounding his distress was the fact that America's security had been compromised.

"You'd have to ask *him*, sir," Hurlburt clipped, lips drawn to a thin line. "But, unfortunately, we don't have time to play who-shot-whom with Paul. You've ordered several B-2 bombers into the air, and they're headed for China. Your decision was made in good faith, based on Paul's recommendation, but Paul's motives are extremely suspect now. Unless we recall those bombers, and do it soon, I guar-an-*damn*-tee they'll bomb the hell out of every ground-based laser site in China! We'll be committing an inexcusable act of war for absolutely bogus reasons! And the entire world will condemn the United States of America for unwarranted aggression. This is as serious as it gets, Pierce. We *must* recall those B-2s immediately, sir!"

Boyer stopped pacing near the straight-backed chairs Hurlburt and Stollach occupied. "I know," he conceded dejectedly, hands jammed deeply into sweatpants pockets. "Paul's been yammering for months about that GBL site—ever since Griffin's spaceplane was zapped. And after the same laser complex attacked that new Lancelot bird—*Damn it all to hell!* Attacking satellites *is* an act of war! I had no recourse *but* to retaliate!"

Hurlburt and Stollach exchanged glances. The president was trying to justify succumbing to Paul Vandergrift's warped arguments and false counsel. This was hardly the time for discourse on *that* subject.

"Ahhh!" Boyer finally waved, resuming his aimless pacing. He stopped briefly, staring into darkness beyond the Oval Office window. The eastern sky was starting to lighten, hinting at a new day's dawn. He turned, arms folded, looking across the office's antique desk.

"All right. Recall the bombers, immediately. I'm ultimately responsible for this damned mess, regardless of Paul's reasons, nefarious or not. I made a mistake. Now we have to undo it, before error becomes tragedy. We cannot let those bombers hit a laser site or a chicken house or a rusty bicycle inside China's borders, understand?" As an afterthought, he asked, "How long before bombs start falling on GBL sites?"

Hurlburt hesitated, mentally calculating. "Not sure. Ground-based laser batteries are scattered throughout China. Once STRAT-COM sends a bomber-release code, then a follow-up confirmation 'go' message, there's no subsequent comm with the birds. Normally, we'd track 'em from space, but that's not feasible, given our current situation. And we don't get routine verbal position reports from crews. That's for security reasons, a holdover from the Cold War, in case communications were knocked out entirely.

"The last we heard from those B-2 crews was an acknowledged receipt of their 'go' codes," Hurlburt concluded, checking his notes. "I'd say we have about four, maybe five hours before the first bomber reaches its target. But remember, sir, there's no guarantee we can reach those birds, thanks to the sorry-assed condition of our strategic-satcom network these days. Especially over North Korea and eastern China."

"Well, our strategic situation has changed dramatically, so you *will* get those bombers turned around," Boyer ordered sharply. "Have General Aster issue recall orders to every one of them. And I'm well aware of the degraded state of our space infrastructure. You've hammered on that point during every morning briefing for months, T.J.! *I get it*, okay?

"Look, this nation's spent a trillion dollars on Cold War-era, nuclear-survivable spacecraft," Boyer said. "A president depends on those gold-plated birds performing, when they absolutely *must* work! Today's that day, so they had better live up to Pentagon hype! And I don't give a hoot about outdated STRATCOM no-comm procedures! Ignore the damned procedures, and get those B-2s turned around! *Now!*" he demanded, his hard eyes betraying—again—a deep-seated suspicion of all things military.

STRATEGIC COMMAND HEADQUARTERS/ COMMAND POST

"Sir, we've got a problem," the three-star director of operations warned.

General Aster looked up from the STRATCOM commander's position on the battle cab's elevated, glass-walled platform. Using a large-screen workstation, he was scanning military-force status reports from across the globe. "What now, Dave?" He had a dull, through-the-temples headache.

"We've transmitted those bomber-recall messages. All but one

B-2 has confirmed receipt of the code and are RTB [returning to base]," Forester said.

"*One* B-2's not responding? Not good! What's the target?" Aster joined Forester at the center of the battle cab.

"That damned GBL site west of Beijing," Forester grimaced. "The one that fired at Lancelot yesterday . . . or today, if you account for the International Date Line."

"The same laser battery that zapped Speed?"

"Same one. The unresponsive B-2 is Spirit One-One out of Okinawa. The crew hit a refueling tanker less than an hour ago, and acknowledged receipt of the initial 'go' code. We subsequently transmitted a confirmation 'go' code, per standard procedures, but did *not* get a 'message-received' answer back from One-One. So, that bomber could be holding and waiting for confirmation, or it could be inside China, headed for Beijing. For whatever reason, it's not responding to recall messages, though."

"Drat!" Aster said, arms folded. "These 'go' code messages typically go through Milstar or the new broadband satellites to strategic bomber crews," he said, thinking through the recall process. "The Milstar constellation is old, but in fair shape, thanks to a nuke-hardened design—*and* being way out in geosynchronous orbit. The Wideband Global System birds are newer, but, for whatever reason, they suffered from the Korean nuke blast. They're not a hundred percent, right?"

"Right, sir," Forester nodded. "We're working with PACOM, trying other comm methods, like UHF and HF radio, but the charged-up ionosphere in that part of the world is really screwing up those options. Bottom line is, we're using everything we have to reach Spirit One-One, but no joy yet." The Army general looked hard at Aster.

"Spill it, Dave. What's on your mind?"

"Sir—just in case we don't get through to One-One, maybe you should give the SecDef a heads-up. If that bomber is already inside China, it could be getting close to the target. If it bombs what China considers a defensive GBL battery just outside Beijing, after we've sent recall . . ."

Aster cut him off. "I'm on the same page. Keep trying to reach that B-2, Dave. You're right. I'd better shoot this up the line. Crap! We told the president this goat-rope of a mission could be disastrous, and now it's headed straight to hell!" The four-star stomped out of the battle cab, slamming the door harder than usual.

Minutes later, Aster closed his communicator's cover, the SecDef's heated words still burning his ears. The news of an out-of-reach B-2 had triggered a classic Hurlburt string of profanity, most of it directed at Paul Vandergrift. Aster put the SecDef's rant out of mind. Only one thing mattered: getting that single B-2 bomber turned around. He reviewed everything he and the STRATCOM battle staff had done so far, looking for that yet-undone option that might spell success.

General "Buzz" Sawyer at USAF Space Command had worked with the 509th Bomb Wing at Whiteman AFB, Missouri, trying to determine whether Spirit 11 might have an equipment problem. An onboard failure would explain why repeated messages sent via satellite communications links were consistently not reaching the bomber. However, the 509th's maintenance group claimed the aircraft's system had been "code one," in working condition, when the bomber took off on its mission to China. Yes, the redundant comm system might have failed inflight, but that was unlikely. Normally, the B-2's comm suite was incredibly reliable.

At Aster's direction, Dave Forester's ops people were frantically trying to contact the B-2 crew via every military frequency available. Time was running out. If it were already inside China, Spirit 11 could be dropping its bombs within the next couple of hours. And all hell would break loose.

B-2 SPIRIT 11 BOMBER/OVER THE NORTH PACIFIC

"No confirmation on that 'go' order yet?" asked Major Edward "Boomer" Booth, a massive young man who dominated the left seat of his bat-winged B-2 bomber. As Spirit 11's pilot, he held the sleek, flying-wing stealth aircraft in a shallow-bank turn, slowly orbiting an imaginary spot in the sky, 30,000 feet above the Pacific Ocean.

"Not yet. Something's playin' hell with the satcom system," the mission commander complained. Shorter, but every bit as muscular as Booth, Major Gustavo "Ringo" Mendoza was a blur in the bomber's right seat. His hands danced between a keyboard on his left and a set of tiny square buttons ringing an instrument panel-mounted multipurpose display unit, or MDU. Also a pilot, he had the additional responsibility of controlling the B-2's navigation and weapons systems, and making critical mission-related decisions. At the moment, his attention was focused on a communications display just above a center console separating the two pilots.

Booth nodded silently, his eyes ranging across multicolored MDU readouts on his own instrument panel. He nudged a fistful of throttles with his left hand, tweaking the power settings of four General Electric F118-GE-100 engines a few percent. He cross-checked engine fan speed against fuel-flow numbers, satisfied that the power plants were sipping as little JP-8 fuel as possible for the current flight conditions.

Ringo Mendoza continued trying various comm channels, repeating what was becoming a well-worn radio call: "Big Pine, Big Pine; Spirit One-One. How copy?" he intoned, enunciating clearly. The two pilots waited, Mendoza's stare trying to coax an answer from the comm system's control panel.

Booth glanced to his left. The Sun had set long ago. Nothing but black sky from horizon to horizon and an endless, now-invisible ocean below. The massive blond pilot shrugged each shoulder, in turn, attempting to shake the tension spreading from his lower back to a bull-like neck.

Now what? he wondered. He mentally scanned the years of training and B-2 flying experience he'd amassed, looking for something that covered the no-confirmation "go-order" situation. What was he missing?

In recent years, the 509th Bomb Wing had gradually moved toward an admittedly vulnerable strategy, one shared by many operational U.S. Air Force, Navy, Marine Corps, and Army units across the globe. The siren call of "space reachback" had led to major changes in B-2 ops plans and command philosophies. It assumed bombers anywhere on Earth would always be tied to their home base or a regional command post via satellite links. After all, there was plenty of redundancy, thanks to multiple spacecraft in orbit.

The Pentagon's logic said: "We'll always have contact with troops in the field, thanks to our robust space infrastructure." Of course, that logic had gone to hell during the 2003 Gulf War, Operation Iraqi Freedom, Booth recalled. He'd been a junior captain then, but had heard the after-action horror stories of Army and Marine Corps units becoming lost and unable to communicate with their headquarters, in the midst of blinding sandstorms. Since then, all such deficiencies had been rectified by an even more-robust space architecture, the 509th had been assured.

Booth shifted his attention back inside. He had flown hundreds of hours with Ringo Mendoza. The guy was an absolute magician, when it came to running all the 'tronics over there in the right seat. If he couldn't get a message to or from the Big Pine command post at Hickam AFB, Hawaii, then something had gone seriously wrong.

As if reading the pilot's mind, Mendoza started a rapid-fire litany. "Our UHF is worthless, thanks to a charged-up ionosphere caused by that damned Korean nuke. We had HF contact for a few seconds, but the ionospheric skip changed after the sun went down. We *should* have solid satcom through Milstar or the wideband birds, but we don't. They're supposed to soak up radiation from a high-altitude nuke blast and keep on humping. They're supposedly the most nuke-hardened spacecraft in history! Why in the hell . . . ?" He rubbed a hand across his neck. Tension was taking a toll on that side of the cockpit, as well.

"Hey, Ringo. Let's think our way through this," Booth said slowly. "We get a coded message that says we're to proceed to our primary target. We've already been briefed that this is a no-shit combat mission, and we're carryin' a bomb-bay-full of precision ordnance. We ask for confirmation, per the book. We get nothing back for a solid hour now, right?" Mendoza nodded, eyeballing the pilot quizzically.

"So, if this were a training mission, I'd say we abort and go back to base in Okinawa. But we're at frappin' war, dude! Those North Korean dum-dums torched-off a nuke at high altitude, right?" Again a nod from Mendoza. "What if the next nuke doesn't go off way to hell up there, but jinks and heads for Seattle? We can't let that happen, man!"

Mendoza agreed, reinforced by a Nomex-and-leather-glove thumbs-up. "I'm with you, Boomer. But we're not going after a North Korean nuke-launch site. Wish to hell we *were* whacking that wing-nut Kim! Oh, no! *Our* target's a mean and deadly ground-based laser site in China, because *that* laser pinged one of our damned satellites!" he grumbled, sarcasm flowing profusely. "Shit, we're not at war with China! And I can't believe we'd start one over the PLA plinking at a stupid *satellite*! It's not like American troops being killed or a bad guy attacking a merchant ship! It's a frappin' *satellite*!" When Ringo got a bug up his tail, the guy was great entertainment, Booth grinned.

"Something's not adding up here," Mendoza continued, waving animatedly, "and I gotta believe some jerk in Washington has his head up and locked, playing chicken with the Chicoms. *We* sure as hell aren't going to kick off World War Three on our own, oh-partner-of-mine! We're supposed to have hard confirmation of that 'go' order, before we depart this fix."

"Yeah, yeah, I know; that's the rule," Booth agreed. "But what if we don't get that order? Then what? And what if that target of ours, the laser site, is one hell of a lot more critical than you and I think it is? Shit, Ringo! We're just a coupla mushroom-majors, floatin' around in the dark, totally ignorant of the big picture! You wanta go home and get your ass court-martialed, because we decided, on our own, that a legal, no-question-about-it 'go' order wasn't real enough for us? We turned back, just because the book says we gotta have a *second* confirming 'go' order? And we made *that* call, on our own, in a time of war? Hell, for all we know, they sent that confirmation and we managed to miss it. Besides, if all the other guys hit *their* targets, we're gonna be *super* dumbasses!"

Mendoza scrutinized Booth silently, considering the options. Night enveloped the bomber, as if they had been swallowed by an empty, black universe.

"Somebody could be jammin' us, too, you know," Mendoza suggested hopefully. "Or something in our comm system crapped out . . ." Finally, he shook his head. "Hell, I don't know. I'll give it one more try. If we don't raise anybody at PACOM this time, we head for the target. Stand by . . ." He punched the transmit button on his control stick, making one last call via satcom, then waited. The distant drone of four powerful fanjet engines and the faint static of radio-silence were the only sounds in the two pilots' helmets. Ringo stared straight ahead through the windscreen, eyeing a sliver of light along the northern horizon. Finally, he jabbed a finger about forty-five degrees to the right.

"Let's go!" he growled, an illusion of confidence covering the sense of aw-shit dread gripping his soul. *What monsters are we about to uncage?* he wondered. Booth carefully rolled the big bomber to wings-level and captured a 320-degree heading. Destination: a nondescript laser facility twenty miles outside Beijing, China.

WATERGATE SUITE/WASHINGTON, D.C.

Paul Vandergrift stared at a plasma-screen TV mounted on the living room's wall, absorbing a flow of information on CNN. The National Security Advisor had faked a nasty head cold, choosing to stay clear of the White House until the B-2 strike missions on those ground-based laser sites were completed. He'd tipped-off a local CNN bureau chief that "something big was about to happen with regard to China." He'd been assured that any breaking news from the cable network's Beijing bureau would be aired immediately.

So far, Vandergrift had been disappointed. By his estimates, the first GBL batteries should have been bombed by now, but CNN hadn't mentioned them. Maybe the damned Chinese were playing "we-know-nothing" again, squelching all news about the strikes. Maybe he should call Feng Bao Nu. Then again, maybe not. If all was going as planned, Nu would have his hands full.

A hint of a smile crossed Vandergrift's features as he took stock of his position. He was witnessing history unfolding, and, in the U.S., only he knew precisely what was happening, and what the inevitable outcome would be.

Iran's neutralized. North Korea's half-blind, thanks to the dumb shits' own nudet. South Korean forces are at the border, but they'll

hold, because an impassioned Pierce Boyer had called Seoul, beg-
ging for restraint, until the U.S. could work out something with
China. Ah, China . . . So far, so good.

He'd taken a huge risk, joining forces with Feng Bao Nu, but
events were playing out as they'd planned. A few more hours and
he, Paul Vandergrift—the future *President* Paul Vandergrift—would
be hailed as one of two peacemakers who had saved the world from
nuclear Armaggedon. He'd already written a few talking points for
his speech to the press; the small card was now secreted in his wal-
let. Not that he needed notes. He'd mentally practiced those press-
conference remarks dozens of times over the last few days.

A flash of uncertainty flickered across his eyes, prompted by a vision of two American aircraft carrier groups patrolling the Northern Pacific. Some of those boats would be sunk within the next
hour or so, assuming the B-2 bombers accomplished their missions
first. As soon as bombs hit several ground-based laser sites, the
Chinese PLA was poised to launch surface-to-surface long-range
missiles from the mainland, zeroing-in on those ships. Although he
hadn't explained, Nu had been confident that the PLA would know
precisely where those carriers would be at zero-hour, enabling accurate missile targeting.

So accurate, in fact, that Nu had assured Vandergrift that neither
aircraft carrier would be damaged by precision-guided warheads on
China's newest antiship missiles. Yes, an Aegis cruiser and a few
other ships in the group would take hits, but that was the price of
being a superpower these days. Your forces were prime targets. Sacrificing a few hundred young sailors was unfortunate, but necessary
to reshape a failed world order, and usher in a period of stability and
peace. A small price to pay for the greater good of America and
China, he'd decided long ago.

Vandergrift's hand shook slightly as he rubbed the stubble on his
chin. He kept seeing mind-movies of ships exploding. He forced the
images away, by mentally retracing the complex twists and turns
that had brought him to this point, a juncture where he and he
alone had made the decision to sacrifice sailors' lives in the name
of a better, safer world. A world in which he would play a much
bigger role, he reminded himself with satisfaction.

From day one, Paul Vandergrift had chafed at his "advisor" role in the Boyer Administration. He was destined for more, much more: a position of real power and authority. To that end, he'd sought and cultivated a number of off-the-books contacts with key Chinese figures over the years.

It had been easy, because his intelligence-community and national-security roles lent themselves to nurturing relationships within the leadership ranks of potential adversaries. Nu, in particular, had proven to be a reliable ally. True, the man was a master manipulator, forcing Vandergrift to trade more than a few tidbits of sensitive information to ensure Nu, in turn, provided accurate, timely insights into the arcane machinations of China's power centers. And Nu had consistently been dead-on correct, which emboldened Vandergrift to eventually propose the plan that was now unfolding half a world away.

Nu had his own reasons for their secret collaboration, of course, but the two men understood each other's hunger for power in their respective circles. And Nu could be trusted. Of that, Van was certain. When the leadership of China and the U.S. shifted during the next few months, a direct result of their now-unfolding plan, Nu and Vandergrift would become heads of state. They would each occupy incredibly powerful positions from which they would align the nations' objectives, guiding two economic giants along a path of unprecedented harmony, peace, and wealth. Of that, he also was absolutely certain.

Pierce Boyer's national security advisor and NSC director stifled a yawn and glanced at his wristwatch. It was already mid-morning. He'd been up all night, too excited to sleep. So much was riding on the next few hours; sleep could wait. As soon as the first news broke about American attacks on China, he'd race back to the White House, despite his "illness." He would jump into the action, nudging and tweaking events here and there in real-time, as necessary.

Where are those son-of-a-bitchin' bombers? he worried, again checking his watch.

17 TRUST BUT VERIFY

GREAT HALL/CHAIRMAN'S CONFERENCE CENTER

Boxed in by military guards, Admiral Lee and Mi Soo were escorted to the Chairman's conference center, a cavernous structure that bespoke power stretching back a thousand years. Lee remembered the facility's splendor from his visits as Ambassador to China. Still, he was awed by the vastness of an arena decorated with ornate furnishings, enormous vases holding live plants and flowers, and works of ancient Chinese art displayed on scattered pedestals.

Quiet grandeur, he reflected. *A monument to human endeavor standing in contrast to a world descending into chaos.*

The building's muted lights and faux walls through the interior created a subtle air of tranquility. Ceilings at different heights corresponding to innocuous partitions formed a maze graced by a soft, feng shui-like aura, enticing visitors to the conference center's core. Tinkling, burbling sounds came from a striking waterfall sheeting across gunmetal-gray slate into black rocks arranged to emulate a creek bed. The rocks glistened like giant pearls. A barely audible lute melody drifted through the room.

The escorts halted short of dividers soaring vertically about twenty feet and arranged in an irregular pattern. Lee and Mi Soo were ushered through a small opening, then a zigzag channel that led to an inner sanctum. Approximately twenty people bustled among the room's workstations. The center pulsed with an air of quiet efficiency. A conference table, seemingly misaligned with the room's layout, dominated the central area. Only three of fourteen chairs were occupied.

Functional. Always functional, Lee noted admiringly.

Wireless headphones and flat-panel screens were at each position. One sported a small, five-inch-tall American flag, reminding Lee of the cheap souvenirs sold by Beijing street vendors. *Made in China, no doubt.*

Mi Soo offered that position to Lee. As he sat, she looked directly into his eyes and whispered, "To avoid war, my trust is with *you,* Admiral." The woman was frightened, Lee realized. He simply nodded.

A large, flat screen on a wall in front of him was divided into three areas. One section displayed a map of China's Far East, Japan, and the Korean peninsula. A second, blanketed with English text, listed the disposition of U.S. forces. Adjacent to each major force were pull-down menus that provided in-depth information about geographical locations. Other sections described the unit's projected intent.

Intent. Hmmm . . . Wargaming inputs? Lee wondered.

The third section was so unexpected that it took Lee a minute to grasp what he was seeing. The screen was devoted to open-source reports about STRATCOM's Deadsats II-Plus wargame, the same venue he had been participating in only days earlier. A "GloCon" Web site hosted by somebody using a "Kosmos" screen identity detailed the ongoing wargame in Omaha. Kosmos had analyzed the wargame's moves and predictions, as had a number of respondents who subsequently added their comments about certain outcomes.

On the vertically split third display, adjacent to Kosmos's GloCon, the Chinese displayed an *Aviation Week & Space Technology* Web site story, which also focused on the Deadsats II-Plus wargame. A reporter named Dane Winfield was writing routine updates to his report, providing personal insights about the wargame's progress. Lee noted that key details were missing from both Winfield's and

Kosmos's reports, because the on-site reporters were not privy to classified elements of the game.

"Admiral! That is from your wargame, is it not?" Mi Soo exclaimed, pointing to the third screen.

"It certainly is," Lee replied, oddly disturbed by the GloCon and *Aviation Week* stories being presented so prominently in a Chinese command and control center.

Why would they brazenly display our wargame's progress? he asked himself. *Perhaps for my benefit, an in-your-face message stating they are closely tracking how STRATCOM deals with satellite losses. Or something else . . . ?*

Lee felt a disorienting sense that he had stepped into an alternate reality. He was in what *looked* much like an American-government situation room, but it wasn't. No, he was inside a beating heart, the decision-making center of a military adversary's command and control facility. On those huge screens, he was looking at American forces through the eyes of a potential enemy trying to predict the Pentagon's next moves. The language and design were functionally different, but the intent was the same. This was a war-planner's room, a strategic hive that could map out the destruction of everything Lee knew and loved. And thousands of miles away, on the plains of Nebraska, warriors in a room much like this one were doing the same. In fact, *this* location, the Great Hall, was probably in those warriors' crosshairs.

Mi Soo slipped into a chair to Lee's left. Three other occupants, seated at one end of the table, ignored the newcomers. They busily manipulated trackballs at their own workstations. Each was wearing a wireless headset. Mi Soo leaned close to Lee and pointed to a second area of the room. "This is what we call a Knowledge Wall. You have similar displays in your wargaming center?"

Before he could answer, his attention was drawn to a flurry of activity. A hush swept across the room as its occupants turned and stood at attention, facing a break in the dividers. Lee immediately recognized the stately figure striding through the opening: Chairman Yi, China's supreme leader. Trailed by an interpreter, the chairman walked directly to the large table.

"Admiral Lee, thank you for coming," Yi said, extending a hand in greeting. "I am most anxious to hear what you have to say. Please forgive me for skipping the usual political pleasantries, but it appears we have very little time," he added, his tone apologetic. "Many currents are driving our ships of state to a precipice from which there will be no turning back."

"Chairman Yi, thank you for seeing me. President Boyer has tasked me to fulfill these official duties on behalf of the United States of America. As his personal representative, I speak for him."

"Yes, yes, Admiral," the chairman said, polite, yet urgent. "I accept your credentials." He waved a palm dismissively. "And the people of China regard your service as former ambassador to our country with great respect."

Lee nodded. The brief accolade signaled the end of political niceties. Critical international negotiations were now under way.

"Our nations are poised for war, and random events are threatening to overcome reasoned thought," Yi declared, gesturing toward the large-screen Knowledge Wall behind him. Icons representing U.S. forces appeared, depicting their assumed state of battle-readiness.

Circling the table, Yi continued, "I fear your mission may have come too late, Admiral. Difficult choices face us, yet we have no time for careful deliberations that could lead to mutual understanding. China and the United States are reduced to preplanned courses of action, and you and I are pawns to unseen, perhaps greater hands. Our pieces, in some respect, are on autopilot, but move with unknown consequences. At the moment, though, I am very concerned about what appears to be an imminent attack—or should I say, *sneak* attack—on my country by an American B-2 stealth bomber!"

Stiffening, Lee wondered how Yi could know about the B-2 missions. The chairman's expression betrayed nothing, but he dramatically jabbed an index finger at the display. A bat-winged B-2 symbol was flashing on the multifunction screen. Originating in Okinawa, a red arc showed the B-2 icon inching steadily toward China's heartland.

Mi Soo rose and whispered, "He wants to know what you know."

"I am aware of the B-2 mission, Mr. Chairman," Lee said with authority. "However, I assure you that mission was a serious mistake, the product of a traitor's deceit. President Boyer has recalled that airplane, but it appears our strategic communication channels are the target of interference. That issue is being addressed, and I can assure you the bomber *will* be stopped very soon. Regardless, I am here to make a very specific offer that underscores our peaceful intents, an offer that can ensure our nations avoid future mistakes, such as this unfortunate B-2 incident."

"Oh, really?" Yi answered skeptically. "Let's see if your President is saying the same, Admiral. I believe he is waiting to join us."

A plasma display descended from the ceiling, halting near the far end of the massive conference table. The American president's image appeared, much larger than actual size.

Lee was visibly disturbed by Boyer's unannounced intervention. The president's presence, even a virtual presence, could disrupt carefully crafted plans. Lee glanced at his chronograph. Timing was critical, and Boyer joining the meeting could make it impossible to implement that plan in the narrow window he would have available.

Maybe the situation has changed—new factors I don't know about, Lee wondered. What would be his role in this new, power-shifted environment?

"President Boyer, Chairman Yi," China's leader began. "I am here with your Admiral Lee. We are discussing the issue of your B-2 stealth bomber that . . ."

"Chairman Yi," the U.S. chief executive interrupted, "this is President Boyer." Lee cringed. Such interruptions were considered very disrespectful in China. Perhaps the inherent delays of satellite or undersea cable communication links were responsible, Lee hoped. Boyer's face reflected deep concern and the stresses of dealing with one international calamity after another. The man looked terrible, his normally handsome features haggard and drawn.

"We are desperately trying to recall that bomber, even as we speak," Boyer said. "It is precisely why I am calling you, because Admiral Lee may not be up to speed on the latest details of how this grave mistake has come about."

Before Lee could respond, a new flurry of activity signaled yet another senior official's arrival. General Luan, commander of the People's Liberation Army, entered, drawing the group's attention.

The tall, slim officer nodded silently to the Chairman, Lee, and the President's larger-than-life image, as he took a seat to Yi's right. The chairman leaned close to Luan, speaking softly in Mandarin. The general tipped his head in respect, before speaking in English, loud enough for the others to hear.

"Mr. Chairman, our courageous air forces are struggling to locate the American bomber." He pointed to a God's-eye-view tactical screen on the Knowledge Wall. "The flight track you see there is only a guess, based on intermittent radar hits. The bomber may be even closer to Beijing." Luan turned to Lee and added, his tone respectful, but cautious. "Many forces are gathering under a darkened sky, and shadow monsters are poised to unleash damning fires."

Cryptic—but I understand, Lee's eyes and hint of a nod answered silently. The Chinese general's expression never changed, but his eyes softened in subtle greeting. Lee and Luan had worked closely, during Lee's years as ambassador, and each man had a healthy respect for the other.

"Admiral, precisely *what* is the B-2's target?" Luan's tone was sharper, underscoring a sense of urgency. "And what other American strike forces are coming? I must know what to defend, what to protect, or there will be no hope of preventing an escalation of hostilities between our nations."

"*No* other offensive actions have been ordered, General," Lee replied forcefully, then turned to Boyer's image. "Mr. President, I must present the details of your proposal to the Chairman and General Luan immediately, *then* we will discuss the B-2 situation. Timing is absolutely critical, as you know."

Not waiting for Boyer's communications-link-delayed reply, Lee approached the Chinese tactical display. Using a small laser pointer, he quickly highlighted several geographical sites inside and near China's borders, outlining the American offer and why it was being made.

"In summary, Chairman Yi, General Luan," Lee said, bowing slightly in their direction, "what we propose is very similar to an offer America made to Chairman Khrushchev more than fifty years ago, when the United States and the Soviet Union also were standing at the brink of nuclear disaster. We now propose a different form of that 'Open Skies' plan. You and your designated representatives will observe the workings of our military apparatus from the inside. By monitoring our real-time planning and operations, you will see that America seeks peace, not war," he concluded, echoing words spoken decades earlier by both President Dwight Eisenhower and Chairman Mao.

Yi, Luan, and their senior staff members, who hovered in the background, were clearly stunned. Their glances reflected disbelief. The American offer obviously far exceeded China's expectations.

The stately retired admiral was in his element, concluding with a theatrical flair and jabbing a knife-edged hand toward the big screen. "Mr. Chairman, our nations have come to the precipice of war for one reason: China has lost *its* situational awareness, or what we call 'SA,' your ability to monitor the movements of American forces. As you may know, my country committed a serious error by temporarily disabling your imaging nanosatellites." Lee noted Luan's jaw tighten, and the general shot an angry look at another PLA officer.

Ahh! You did not *know what happened to those nanosats, eh, General?* Lee flashed with satisfaction. "But *your* loss of SA was compounded by the North Koreans' very unexpected detonation of a nuclear weapon in the upper atmosphere, which disabled many of *our* satellites," Lee continued. "That was an unprecedented act of war, not only against us, but against you, a North Korean ally. Their misguided act seriously degraded the operation of your, and our, space-based assets. However, America is quickly reconstituting some of its on-orbit situational awareness."

Lee allowed that declaration to hang in the air a long beat, its implications clear. Had China controlled the psychotic North Korean leader, space-based monitoring systems would still be in place and there would be no misunderstanding now—Vandergrift's and Nu's meddling notwithstanding.

"To compensate for *your* loss of SA, the United States is offering *its* situational awareness resources to China, including highly classified imagery. To reiterate: We are offering to share our most-prized intelligence capabilities with China's civil and military leaders. To preclude a war, America will provide you with full insight to our operations in the Pacific region. Such visibility will prove our peaceful intents. Mr. Chairman, we will reveal to your representatives the exact locations of our forces in the Pacific, as well as their readiness posture, all in real-time."

Lee quickly listed the reconnaissance and surveillance assets the U.S. was making available immediately. A number of so-called "spy" aircraft equipped with powerful radars and electro-optical systems were standing by at air bases in Japan and South Korea. Chinese PLA delegations would be invited to board them for a first-hand look at the intelligence data they gathered. Further, real-time space-based intelligence, although degraded at the moment, would be available for viewing at the American embassy in Beijing.

"Mister . . . *Chairman* Yi," President Boyer injected. "You should know that the offer Admiral Lee is presenting was not an easy sell here in the United Sates. I mean . . . with the political situation we face. I'm sure you understand, in your position . . . Despite that, we are making every effort to assure China that our intentions are peaceful, even at our own considerable risk."

Lee squirmed, silently grimacing at the President's visible unease. *Damn! Airing concerns about domestic political bull hardly projects a powerful image of American leadership!*

Lee stepped toward the chairman, intervening before Boyer exposed a weakness that Yi could easily convert to a Chinese victory. "Mr. Chairman, we know you, too, face a serious dilemma. We are aware of elements within the Party seeking to amass power during these unfortunate times. They do *not* have China's best interests in mind." He left the vague charges hanging, unexplained. Yi's demeanor was unreadable, his eyes locked with Lee's.

Luan raised an eyebrow and stole a glance at the Chairman, before responding. "Yes, Admiral. We *do* face internal difficulties."

"If we cannot stop our B-2, and it completes its very unfortunate, mistaken strike mission, what will be your response?" Lee asked bluntly. Time was slipping away.

"Mistaken? *Unfortunate*, you say? I doubt that will be the perception of our people, Admiral!" Luan replied with equal candor, his English crisp and proper. "If the B-2 bombs fall on Chinese soil, the matter will be out of our control. The wrong faction, the forces you refer to, will have an excuse—and the authority—to carry out war plans already in place. Of course, China will respond! We have long-range forces poised to strike back automatically. They will not await *my* orders!"

"The 'wrong faction,' General? You refer to Feng Bao Nu, correct? What is his status, his power base?" Lee shot back, speaking rapidly. He was pushing the envelope of political civility, but he spotted an opening, the inside wrist of an opponent, and, like the expert fencer he had once been, he thrust. He must learn what the Chinese leaders knew about the Vandergrift-Nu pact.

"Admiral Lee, *I* will address that issue," warned Boyer sternly.

"Mr. President," an increasingly impatient Yi barked, "*that issue* must wait! Let us return to the heart of a matter at our throats now. Your errant B-2 threatens to tip a delicate balance. A strike on China will be seen as a serious act of unprovoked aggression, and all reasonable voices will be drowned in the clamor for retaliation. The shadow forces Admiral Lee refers to will drive mobs into the street, demanding American blood. You expect me to control them, simply by *apologizing* for your mistake? Impossible! At this moment, even *I* must suspect your *full* intent!" Yi, too, was probing, using the impending strike as leverage. Lee hoped Boyer saw through the chairman's ploy.

Boyer walked around his desk, approaching the camera until his form filled the Chinese center's oversized screen.

"Mr. Chairman," the U.S. president began, his demeanor shifting to that of commander in chief, the military leader of Earth's remaining superpower. "There *is* no other intent! My *only* intent is to stop this fool's rush to war! You must know that my own National Security Advisor, Paul Vandergrift, has cultivated an unauthorized liaison with your Feng Bao Nu, the man Admiral Lee just mentioned. This is not an accusation of Chinese misdeeds, sir. We have undeniable proof that Vandergrift and Nu have their own, private ambitions and forged an extremely dangerous devil's plan that will ultimately compromise the best interests of our two nations. Vandergrift is operating without my concurrence, and I consider his actions high treason against America. I suspect that Nu . . ."

Luan nearly leaped toward Yi, startling Boyer in mid-sentence.

The Chinese general whispered something to the chairman. Yi nodded and apparently gave Luan an order, before walking to a nearby communications station.

Yi's movements served to bring him closer to the videoconference camera, as well. Images of both Boyer and Yi now filled their respective video frames, elevated tensions starkly visible in the hard-set of each man's face. Lee had seen those looks all too often during combat operations. A simple miscalculation and people would die.

How many souls this time? the retired naval officer wondered, a wave of hopeless weariness flowing into his being. He shook his head slightly, trying to clear it. If ever he needed all his considerable mental faculties, this was the time.

Yi matched Boyer's serious, commanding air. "We suspected Nu was dealing with someone in your government. However, I am surprised at your revelation, Mr. President. I have just now ordered General Luan to find and detain Feng Bao Nu for interrogation. I would hope you do the same with your security advisor. Meanwhile, how can you possibly guarantee that your intent, indeed, is peaceful?"

"Mr. Chairman, really," President Boyer began. "The problem at hand . . ."

But Yi cut him off, even before Lee could. "*This* is the problem at hand, Mr. President: *your intentions!*" Yi shouted, his words ringing with passion. "You think the machinations of your political system are invisible to the rest of the world? You break off talks with the Korean People's Republic to force them into submission. Your two carrier groups in the Pacific frighten the North, leading Kim to detonate a nuclear device, just to blind you, trying to prevent an invasion of his country. And now *you* complain? America always needs enemies for political fodder, Mr. President. Only this time, if that bomber reaches its target, what you call a misunderstanding will certainly wrest control from your hands and mine. This dangerous game will end and we will be at war."

"What my predecessors did, Mr. Chairman," Boyer began, his voice rising to discourage another interruption, "I cannot change at this moment. I cannot control *your* Mr. Nu's previous back-channel deals with *my* Mr. Vandergrift. I will take very firm measures with

Vandergrift, I assure you. But, for now, you and I can only shape what happens in the next few hours. Please set aside your concerns about America's long-term intentions and help me stop that B-2. First, please immediately cease your jamming of our satellites' communication links!"

"Your true intentions, Mr. President, were implicit in how your nation chose China to be your adversary," Yi continued, ignoring Boyer's plea. "For twenty years, your military strategists and other influential thinkers have labeled us, China, as the '*new enemy.*' Is Admiral Lee's mission nothing but another feint, while the final thrust into your '*new enemy's*' heart is yet to come?" He shot a hard, questioning glance at Lee. "Are you buying time to guarantee your stealth-bomber strikes will destroy my people, my nation? *How can I know?*" he shouted.

And the B-2 draws ever closer, Lee thought, despairing. The two leaders paused, leaving a strained silence of desperation hanging between them. In that instant, Lee decided his moment had come, the time to bring all his skills to the forefront. A lifetime of learning, experience, judgment, and combat-honed savvy was demanded. Deep inside, calm and confidence ruled. *I have not yet begun to fight*, Lee grimly declared to himself. He stood and deliberately approached Yi, carrying an aluminum-sided case.

"There is more we will provide, Mr. Chairman," he said, radiating determined strength. "Right here, right now." He glanced at Boyer, but read nothing in the man's stern, oversized screen image.

"Time is of absolute essence, Mr. Chairman," Lee stressed again, patting the case. "This is a compact mobile command and control system. It provides the same real-time imagery and classified data that Strategic Command, the American Joint Task Force Commander in Hawaii, and field commanders in charge of maneuvering thousands of troops, aircraft, and ships are viewing right now."

Luan, who had slipped back into the group now gathering around Lee, sensed the admiral's strategy. The Chinese general interjected a few comments to Yi, amplifying on what Lee was explaining. Although Lee could generally follow the Mandarin dialect, much of the quiet, rapid-fire dialogue between Luan and Yi slipped by.

Yi finally shook his head, raised both palms as he turned from Lee and faced Boyer's image. "Mr. President, Admiral. Why is it you can instantly show me the location of American forces throughout the Asia-Pacific region, yet you cannot recall one bomber from inside China? I find that difficult to reconcile! We have no time for looking at pictures!" Yi declared.

"Feng Bao Nu is now detained," he added, "but I still do not know whether your misguided B-2 is the vanguard of a devastating nuclear strike, or a clever ruse to force a retaliatory response. The latter could kill thousands of your sailors and give you reason to unleash your nuclear arsenal. Yes, your offer to share America's most-secret intelligence is impressive, but I fear it is too late."

Lee pressed on, choosing to ignore Yi's bullheadedness. "Chairman Yi, General Luan. I concede that there is no time to deploy China's representatives to our command centers and inbound aircraft, even though that would provide physical, eyes-on confirmation of America's current strategic posture. But we can still deliver on our offer, because, with this equipment, I can restore China's situational awareness. Armed with proof of America's true intentions and military posture, *you* will hold the confidence of your Party's leaders. And General Luan, not Nu's comrades, will remain the undisputable leader of the PLA throughout very critical days to come."

Lee carefully positioned the thick aluminum case on the conference table and opened it, revealing a compact digital communications system. "As for our aircraft carriers," Lee continued, "this system will show you exactly the same overhead imagery and radar data that our intelligence analysts are receiving. You will see that our naval forces are *not* in a warlike posture."

He lifted a folded panel, set it on tripod legs and adjusted the panel until it pointed skyward. The laptop computer's screen came to life, asking for "Secure Log-In" information. Yi, Luan, and Mi Soo stood behind Lee as he entered an access code, then waited while an image painted the screen, top to bottom. The flat panel tilted slightly, reorienting itself via integral servo-drivers built into the tripod.

"This system automatically configures itself to receive the data stream I've requested," Lee explained, intentionally omitting the fact that those data were encrypted. "This is a fused image consisting of visible-light, infrared, and radar data acquired periodically by two new American satellites in a highly elliptical orbit. You're now observing an air base in Okinawa. A number of our aircraft were launched from there, when the North Koreans threatened to launch yet another nuclear weapon." *As was that damned B-2 bomber headed our way*, Lee thought, jaws tightening. Time was definitely running out . . .

Yi leaned over Lee's shoulder, tilting his head back to better view the real-time display through bifocal lenses. Lee continued, "These images are being beamed down to us from Arthur, a new American satellite now passing over China. Its sister spacecraft, Lancelot, was attacked only yesterday by a ground-based laser site near here." Lee shot a hard glance at Luan, who pointedly focused on the computer's compact screen.

"Admiral, this is all very interesting, but please excuse my—skepticism," Yi interrupted quietly, yet firmly. "Could not these images simply be prerecorded on this system? Or, perhaps, manufactured by your Pacific Command headquarters, a . . ." He paused, searching for just the right words. ". . . movie created only for our benefit, to trick me? I have no assurance these are little more than a digital simulation!" He started to turn away.

Lee clipped, "Chairman Yi, I understand your skepticism. In your position, I would have the same doubts." He tore a page from a yellow legal pad lying on the table, the noise pulling Yi's attention back. "To prove that we are sharing real-time images transmitted from some of America's most-secret intelligence platforms, I suggest a test: Please send one of your assistants outside this building and have him place this sheet of paper on the ground, visible from the southern sky." Lee handed the page to Yi, then added, "Please write something on the sheet. In large characters, preferably." He met the chairman's glance and smiled. Both knew how powerful U.S. optical satellites were purported to be.

Yi accepted a thick-tipped marker from Mi Soo, bent and scrawled several Chinese characters on the yellow sheet. Lee intentionally kept his eyes averted, focusing on the small computer screen as it panned along a flight line in Okinawa. *Several empty slots,* he noted. A few hours ago, there had been a dozen or more B-1 and B-2 bombers parked there.

Yi handed the sheet to one of his assistants and barked an order. Before he could depart, Lee grabbed the man's arm and gave him a six-inch-long infrared flashlight, explaining what to do with it and the sheet he held. The assistant broke into a run and disappeared behind a partition.

The chairman looked at Lee, pointedly slipping a personal communicator earpiece and microphone into place. *He'll be in direct contact with that man,* Lee thought, satisfied. The dog-and-pony demo was going precisely as expected. From the other end of the table, Boyer's extra-large image silently watched the activities from Washington.

"While we are waiting, may I make a request, Admiral?" Luan asked, more a command than question.

"Of course, General. Your request . . . ?" Lee's hands were poised over the computer's keyboard.

"May I see the two American carrier groups that are steaming toward our shores? I would like to personally assess their combat readiness. I'm sure you understand," Luan said, firmly.

"Absolutely," Lee nodded, tapping several keys and dragging a fingertip across a touch-sensitive pad to access a series of pull-down menus. He leaned back, enabling the small group to view an image that popped up.

"This is coming from the new Arthur platform, which is flying over China now. Because of its relatively low altitude and elliptical orbit, please understand that Arthur will only be available to us for a short time. Then it will be out of range for hours. We recently deployed it and the Lancelot spacecraft to ensure our commanders have reliable contact with forward-deployed forces, such as these carrier groups. The North Korean detonation destroyed several American satellites that would normally ensure positive control of our deployed forces. Correction: I should say the high-altitude detonation destroyed the *electronics* payloads on those satellites," he added, without further explanation.

Luan studied the image, then asked that the image be expanded. At one point, Lee pointedly swept the camera along a carrier flight deck, clearly revealing the subdued name, *Lincoln,* emblazoned on the gray steel. A line of F/A-18E/F fighters was parked at an angle along one edge of the deck. All but two aircraft had their wings folded and were firmly tied down.

"Those two Hornets. Alert aircraft?" Luan asked, pointing.

"Yes. But you'll notice the others . . . ," Lee began, but was interrupted.

". . . are all secured," the Chinese general finished. "Are you in contact with that ship, Admiral?"

"I can be," Lee said, pulling his communicator from an inside jacket pocket. "Would you like to talk to the captain of that carrier?"

Luan smiled and tilted his head, the picture of doubt. "I certainly would."

Lee returned the smile. "Then I must ask for your assistance, as well, General. If you would be so kind as to have your Army cease its jamming of our Milstar satellite's downlink, I can call that carrier immediately. Then you can talk to the captain, if you'd like."

Luan's dark eyes hardened, boring into Lee's. The admiral held the other's stare. Neither blinked. Two warriors toe-to-toe, each waiting for the other to flinch. Luan slowly shook his head.

"My apologies, Admiral, but I cannot do that, because we are not jamming Milstar or WGS or any other American satellite. I'm sure you know that those spacecraft employ spread-spectrum, frequency-hopping and other anti-jam techniques, making their signals virtually impossible to jam from a distance. Have your General Aster look elsewhere. China is *not* jamming your satcom!"

Lee held the general's hard, steady gaze for long, tension-filled seconds. He finally nodded curtly. *On to 'Plan B.'*

"Very well, General. In that case, I cannot guarantee we can reach the *Lincoln's* captain, but I will try to call him through the same Arthur minisatellite," Lee explained. He reconfigured a couple of transmit protocols on his handheld communicator's dropdown menu. *This'll be tricky,* he thought, trying to will away a flood of doubt. Arthur's comm package simply didn't have the power and range to reach across excessively long distances. And would its signals also penetrate the concrete and steel of the Great Hall?

If this works, will the message even get through in time? he wondered, throat tightening. Each passing moment was his enemy. But these leaders also realized that.

Yi spoke into his communicator, pressing the earpiece closer to hear better. He caught Lee's attention, saying, "My assistant is outside, Admiral. At the south portico, on the steps. He has placed the paper on the step, as you requested." The chairman's lips were a thin line, doubt permeating his body language.

Lee glanced at Luan, who nodded sharply. The American turned back to his computer, noting Mi Soo's worried look. He winked and was rewarded by the young woman's nervous, brief smile. Working quickly, Lee selected a series of menus, opening a database now being constantly refreshed via Arthur's downlink. Not as ideal as if he were tapping the data from more powerful, versatile "national technical means" intelligence-gathering spacecraft, but the mini-satellite's limited capabilities would have to do.

This had better work, he thought grimly.

An image appeared, clearly showing a false-color top view of the Great Hall, despite the darkness that now enveloped Beijing. Lee commanded a closer look, then selected an area along the building's southern periphery. Activating a "lensing" feature, he zoomed even closer within the area's confines, leaving the rest of the image untouched. It was a clever feature that enabled a viewer to maintain big-picture situational awareness, while closely scrutinizing a particular part of the larger image.

There! A figure stood on broad steps, next to a rectangular object brightly illuminated by a pool of infrared light. Lee smoothly, continuously zoomed toward the object, then expanded the area-of-interest until that object filled the computer's image. He smiled. *Son-of-a . . . That's incredible!* he thought, astonished. Sure, he'd been briefed, but to actually see firsthand an image created by fusing infrared and visible light with side-looking radar signals . . .

The chairman was equally impressed. "Admiral, I am convinced. I can read the paper, and that *is* what I had written. Very impressive, indeed." Thanks to the infrared flashlight held by the assistant, Lee also could read: "Chairman Yi," in bold Chinese script.

General Luan cleared his throat, subtly suggesting Lee resume his attempt to communicate with the carrier group. The admiral quickly said, "Chairman Yi, with your permission, I would like to show General Luan something." Yi flicked a hand, his smile quickly fading.

"Admiral Lee, need I remind you that the B-2 bomber is *still* out there? Please hurry . . ."

Lee nodded, rechecking the handheld communicator's configuration. He tapped the screen with a stylus and touched a "Send" icon. While a uniquely coded signal raced up to the satellite and down to the *Lincoln*, Lee selected the device's speaker mode, enabling the others to hear a response. No more than two seconds passed, before a deep voice answered.

"Captain Henley here. How can I help you, Admiral?" Lee smiled to himself, gratified a comm link had been established prior to Arthur slipping over the horizon. He quickly explained the situation to the commanding officer. "And please do whatever General Luan asks."

"Aye, aye, sir," Captain Henley snapped.

Lee handed the communicator to Luan.

"Captain, this is General Luan of the PLA," the tall officer said, leaning toward Lee's computer screen. "I am looking at an image of your flight deck. Admiral Lee assures me that this is what you call a 'real-time' picture. To convince me, I would appreciate having the canopy of your number-one alert Hornet lowered, then raised. I will be watching."

"Certainly, General. Stand by, please." Luan heard a muffled command as the captain shielded his communicator's mouthpiece. Luan pointed at the laptop screen's image of the F/A-18 he'd designated. Lee nodded and zoomed the image closer, halting when the fighter's forward fuselage filled the screen. The helmeted cranium of a pilot lounging in the cockpit, one foot propped on a side console, was clearly visible.

The canopy of that Hornet slowly closed. A second later, it opened, halting at the former full-open position. The helmeted pilot looked up and swung his arm side-to-side. Luan laughed and muttered something indecipherable to Lee.

"Admiral, you have convinced us," Chairman Yi declared. "I must say, I never thought America would share such highly secret information with my nation. You are an honorable man. I salute you—and your president," Yi said, louder and tilting his head toward Boyer's image.

Yi added, his tone sharper, "President Boyer, I am prepared to stand down our nuclear forces, based on the good faith you have shown with Admiral Lee's impressive demonstration. I am satisfied that we now can monitor your carriers' position and readiness states. However, I cannot leave China defenseless. Until we know your B-2 bomber is no longer a threat to my nation, I will keep our conventional forces on high alert. Now, how do we stop your B-2 bomber?" he asked, eyes blazing again.

B-2 COCKPIT/OVER CHINA

"Any bad guys out there?" Boomer Booth asked casually, gently banking the bat-wing B-2 bomber to pick up the new heading Ringo Mendoza, the mission commander, had designated. The aircraft's ingress route had been planned to skirt known Chinese air defenses. Still, the pilot concentrated on making careful control movements to limit deflections of the bomber's barn-door-like drag rudders. When open, those huge control surfaces could momentarily reflect radar pulses back to powerful Chinese search radars on the ground. Detection of an unknown intruder headed straight for Beijing would almost assuredly prompt the PLA to scramble fighters. Hell, fighter-interceptors could be in the air already, guided by radar operators searching their screens for *any* sign of the B-2.

Fighters worried Booth. Damp spots under each armpit were steadily spreading, and the pilot kept wiping tiny trickles of sweat running from beneath his helmet and across his forehead.

"You know, we probably have half the frickin' Chinese air force looking for us, trying to shoot us down," he commented quietly. Mendoza didn't respond, his fingertips tapping keypads arrayed on the center console and forward instrument panel. Cross-checking a number of settings, he then blasted a very short-lived burst of low-probability-of-intercept radar pulses into the night, while eyeballing a color display. A few microprocessor-generated blips appeared, then turned red.

"Four fighters along the east coast, headed north. No factor at this time," Mendoza clipped, pointing to a colored map on one screen. "If things turn to shit, after we make the target run, we'll swing east, duck behind those bad guys and make a run for the water."

Assuming those fighters don't get a tally and turn on us first, he didn't have to add. He knew Boomer was thinking the same thing. Their B-2 was deep into enemy territory, the target only minutes in front of them, and home-base was an hours-long flight away. *We're all alone, waaaay out on the frayed end of a damned long string,* Mendoza thought, trying to shake a sickening sense of dread.

GREAT HALL/BEIJING, CHINA

Visibly shaken by Yi's unvarnished challenge, President Pierce Boyer was incapable of responding immediately. That, plus Mi Soo's subtle body language and eye contact, reinforced Lee's conclusion: This was *the* pivotal moment. He suddenly realized that his entire life had been leading up to this moment. This was Stanton Lee's reason for being. This was his hour to forget careful planning and go for dramatic intervention.

Since his first days at Annapolis, Lee had privately dreamed of boarding a British frigate amidst cannon fire and musket rounds, hacking his way through netting, and standing sword-to-sword with a young Horatio Hornblower on the pitching quarterdeck. In a way, this, indeed, was that moment of dreams. Sure, he'd been a fighter pilot and had watched the arc of deadly surface-to-air missiles, as they tracked his F-4 Phantom above the Ho Chi Minh Trail. But at heart, he was a fighting sailor, standing with John Paul Jones as the Ranger sailed into enemy waters.

A flashing icon in the upper right corner of Lee's compact computer screen caught the man's attention. While Boyer and Yi traded barbs about the errant B-2 bomber, Lee clicked on the icon. A very short, curt message from General Howard Aster popped up. Lee scanned it quickly, alarmed at first, then resigned.

However, in a flash of ultra-clear decisiveness, the clouds of doubt parted and Lee's destiny shone as rays of brilliant sunlight slicing through the salt spray of a sparkling ocean. He was infused with the power of instant knowing, the thrill of absolute certainty. In a flash of supreme acceptance, he mentally and emotionally stepped forward and embraced destiny.

Unobtrusively, Lee quickly typed a few sentences, pulled down a menu and selected *"Sleep."* The screen blanked, but a tiny light-emitting diode above the top line of keys continued to blink, indicating the computer was still operating.

18 NIGHT SPREAD OUT BENEATH THE SKY

STRATCOM HEADQUARTERS/COMMANDER'S OFFICE

Lieutenant Colonel Burner Burns couldn't help overhearing General Aster's terse three-way conversation with Space Command's General Buzz Sawyer and Aster's four-star Pacific Command counterpart in Hawaii. Earlier, none of Space Command's or PACOM's space-smart wizards had been able to determine why the Milstar and Wideband Global System satellites' encrypted communications downlinks were ineffective in reaching Spirit 11. They'd finally given up and were searching frantically for other comm options. The B-2 was still inbound to China, hell-bent to bomb that ground-based laser site, its crew oblivious to the yeoman efforts being made to get a recall message through to it.

Burns turned back to the pile of yet-to-be-signed memos in front of him, but something kept nagging at the edge of his concentration, tickling, coaxing his mind. Something important. Irritated, he shifted uncomfortably, his toe brushing the aluminum "football" on the floor. He glanced at the case, again sensing a wave of ill-defined unease. *What the hell . . . ?* he wondered, frustration gnawing at his gut.

Taking a deep breath, he resorted to a private, very personal tactic that had worked well throughout his stellar flying career. *Clear the mind. Let it float. Don't force the conscious mind down a particular pathway.*

He stared at a paneled wall in Aster's office, letting the general's relentless pacing and intense phone conversation drift to the periphery of awareness. Burns freed his mind to wander across time and space, transiting thousands of miles, seeking that single B-2 bomber now cruising toward China, a black specter of power and destruction cloaked in a night sky.

He knew the Spirit 11 pilots. For several years, Burns had served as operations officer for the famed 509th, the nation's only B-2 wing at Whiteman AFB, Missouri. Consequently, he knew every man and woman who flew the sophisticated stealth bombers. In fact, he'd been personally involved in selecting both Mendoza and Booth for B-2 duty. He'd also given Ringo Mendoza a checkride that upgraded the blocky Hispanic to Mission Commander status.

Both were excellent pilots and dedicated officers. They *would* complete their mission and destroy that GBL site. Of that he was certain. Astoundingly, the U.S. commander in chief had ordered that any and all measures be taken to stop them *before* they succeeded in that mission. In this convoluted, bizarre moment, Spirit 11's success would spell abject failure for America, opening a door to even more chaos and, potentially, destruction on a horrific scale.

Burns let his mind wander about the bomber's familiar cockpit, taking note of the four large, flat-screen, color displays on the instrument panel in front of each pilot; the button-encrusted control stick between each man's thighs; the bank of engine throttles at each of their left hands; the few round-dial standby instruments that seemed out of place in such a modern cockpit, and the forest of press-to-activate square buttons blanketing a broad console between the two crewmen.

Was it just memory, or was he actually "seeing" those very objects via some yet-undiscovered means of remotely viewing Spirit 11's flight deck? Finally, an area jumped at Burns, attracting his consciousness; a small, familiar device clipped to the edge of a zippered flight suit slash-pocket. It was Boomer Booth's flight suit pocket. In an explosion of insight, Burns had the answer his boss was seeking.

The executive officer jumped to his feet and approached Aster, trying to attract the general's attention. Still on the phone, Aster was staring out the office window, nodding, and listening to yet another option raised by one of the other four-stars. Finally, the STRATCOM chief noticed Burns and raised one eyebrow in a semi-irritated silent question. Burns stage-whispered, "I've got it, sir. I know how to reach that B-2's crew!"

Aster's forehead wrinkled, not understanding. He mouthed "What?"

"I can reach that crew—the B-2 that's headed for China," Burns repeated quietly. "But we need to move *now*, sir!" Aster told the other two four-stars he'd get back to them, then faced the lieutenant colonel.

"Burner, what the hell are you sputtering?"

"Sir, it just occurred to me. Every B-2 pilot at Whiteman carries a satphone communicator that doubles as an all-in-one cell phone, personal data assistant, video player, whatever. One of our guys found a hot deal that offered a military price break, so we all bought the new handhelds that hit the market when the Excalibur com-sats were launched a few years ago," Burns said, speaking rapidly. "They're dual-purpose, a blend of conventional cell phones and sat-phones, but packaged as a single communicator. They allow crews to stay in touch with their families and squadrons, even during over-seas deployments. Hell, we used 'em as backup operational comms when we couldn't get priority on a milsat link for some reason. Not exactly opsec-approved, but . . ."

"Operations security damn-sure *would* be compromised," Aster growled, hands on hips, looking down at Burns. "Still . . . this could be exactly what we need to contact that crew!" he declared, smack-ing a fist into a palm. Then an "aw-shit" flashed across his chiseled features.

"Damn it, Burner! That North Korean nuke killed every low-orbit commercial comsat in the Pacific region," Aster said, relief turning to renewed frustration in a heartbeat. "Damned few Excaliburs are still operational!"

"Maybe there's a way around that, sir. I saw a flash message come across a few minutes ago, indicating the commercial companies have been scrambling to restore comsat service," Burns said, shuffling the papers he carried, looking for a brief missive. "Pantera, the outfit that flies the Excaliburs, has, quote, 'robust recovery plans in place, including on-orbit spares, power-restoration techniques, and graceful-degradation features,' unquote. Because they guarantee their customers a high level of service, they're motivated to restore full-spectrum service."

"Yeah, but the Excalibur constellation was already hurting! That damned druggie's maser took several of those birds down. And now this nuke . . ."

"If I may, sir," Burns interrupted firmly. "We have a Pantera guy downstairs, working with our wargamers. He knows the Excalibur system cold. Damned sharp guy, so maybe . . ."

Aster snapped, "Right! Joaquin!" Raising his voice, the general called to his assistant. "Annie! Have Jim Androsin grab what's-his-name Joaquin, the Pantera rep, and get him up here *ASAP*! Tell him I mean *now*!" Turning back to Burns, he continued. "Okay, then what?"

"Sir, if Joaquin can show us where and when we have Excalibur coverage over the Pacific, maybe we can slam a phone call through to that B-2 crew," Burns said, pulling a slim, palm-size communicator from a plastic holder attached to his belt. "A long shot, but it may be the only shot we'll get."

Aster nodded, reaching again for the phone on his desk. "Rog! I'll have Annie get the pilots' contact info from Whiteman."

"Sir, no need for that. I've got their numbers right here. I had all the B-2 pilots' numbers programmed into my 'Sword' communicator, when I was the D.O. at Whiteman," Burns said, tapping the communicator's small, touch-sensitive color screen. "Yeah! Still got 'em! Now, if those SOBs just have their handhelds turned on . . ." Burns inserted a tiny wireless earpiece in his right ear and punched a transmit button, initiating the call. The earpiece glowed a bright blue, indicating activity.

Aster and his executive officer waited silently a few seconds. "Shit! 'No service available!'" Burns swore, clicking the device shut. "Maybe you're right, sir. Maybe there *aren't* any functional Excaliburs out there." He grimaced in disappointment.

Colonel Androsin and Rob Joaquin appeared at the door of Aster's office, prompting the general to wave them in impatiently. He quickly outlined the problem and Burns's just-failed attempt to call the B-2 crew. Joaquin stepped forward, a hand raised to interrupt Aster's monologue.

"General, we might still be able to get a call through. Can I use your computer?" he asked, already moving to the large-screen terminal on an extension running perpendicular to the commander's oversized, polished desk.

The other three clustered around Joaquin, watching the Pantera rep's fingers fly across the keyboard. "I'm logging into the Excalibur satellite control center," he explained, pointing at a graphic as it flashed on-screen. "I want to know which birds are operational out there in the Pacific. Some were over the horizon and survived that North Korean nuke's EMP. And our guys are bringing more of the standbys online every day. We still *do* have holes in coverage, though. . . ." he cautioned, voice trailing off, as he focused on manipulating the on-screen image.

"There! See that satellite, just clearing the horizon? It'll cover a good-sized chunk of eastern China in a few minutes. But we'll only have a brief window to get a call through, because that bird's coverage area will sweep across eastern China mighty fast. It's not at an optimum angle, either," Joaquin warned, pointing to the screen. A satellite icon was just visible above a depiction of the Earth's curved horizon, its fanlike beam of coverage sweeping across the globe like a push broom.

"Usually, we have multiple spacecraft over any particular region, and they give us overlapping, constant coverage. But we don't have a full constellation anymore. Colonel, what type of comm system are you using?" Joaquin asked, glancing at Burns.

Burns extended the communicator toward the Pantera expert. "Sword Twenty-Four," the officer said.

"Not as powerful as the new 'Mace' series. It oughta get through," Joaquin declared, accepting the device. He tapped the screen rapidly, searching for something buried in arcane lines of code—a company-provided key that allowed Burner's device to access a certain remote communications network. He found it, pointed the device's top edge at Aster's computer workstation, and tapped the communicator's screen again. The displays on both machines flashed blue for a brief second.

"Done," he said. "The handheld and this computer are communicating. Now, please stand over by that window and be ready to tap this thing when I give you a 'go' call," Joaquin ordered, handing Burns the device. Turning back to the computer, Joaquin adjusted the computer screen's graphic, expanding the Excalibur's "painted" area of the Earth. "Do we know where that B-2 is?"

"Not for sure. Probably somewhere over the southeastern China mainland," Burner said, stepping to the window. "Ready when you are," he announced, making sure a ten-digit number was visible on the communicator's small screen.

"When you get ahold of one of those pilots, I wanta talk to him ASAP," Aster demanded, red-rimmed eyes boring into those of his executive officer. Burns nodded, then focused on the PDA-like device.

"Ready—*now!*" Joaquin commanded.

Burns tapped the screen and adjusted his earpiece. His eyes lit up and a crooked grin started to spread. He handed the device to Aster. "Sir, it's ringing. Major Booth should answer, assuming he has his comm system turned on." *The bastard* better *have it on!*

Aster, not waiting for the wireless bud, put the device to his ear and waited. He finally glanced at Burns and nodded. Someone in the B-2 was answering.

"Major Booth!" the general cried. "Thank God! Major, this is General Howard Aster, commander of Strategic Command in Omaha. I'm ordering you to scrub your mission. Repeat. You are to *abort* your mission immediately and return to base. You copy?"

Aster waited a long moment before he heard Booth respond, "Sir, do you have a recall code for us?"

"Rog. Stand by. We'll have the code in a second," Aster pointed to the "football" Burns had already retrieved. The lieutenant colonel quickly cracked the aluminum case, grabbed a plastic-covered booklet, and thumbed to a page before handing it to Aster.

"Guys, better hurry or we're going to lose him!" Joaquin warned, eyes locked onto the computer screen's animated graphic. "That Excalibur's signal is barely touching eastern China."

"Major Booth, listen up. We have a very short window before an Excalibur satellite transitions your coverage area. If we lose contact, you *will* turn that bomber around instantly and beat-feet back to base! That's an order issued by the President of the United States. How copy?" Aster said, voice now hard and commanding.

Across the dim cockpit of the B-2, Booth raised a skeptical eyebrow at Mendoza. Both were listening to Aster via speakerphone on the pilot's handheld communicator. Mendoza frowned and mouthed the word, "Code!"

"Sir, we need that authentication code," Booth said firmly.

Aster grabbed the booklet from Burns and read a long series of alphanumerics. Thousands of miles away, Mendoza carefully recorded the string of letters and digits, then compared it to a small display on another handheld device. Part of a new "electronic flight bag" system all B-2 crews carried, the battery- and solar-powered system had displaced the old paper code books lugged around by strategic bomber crews decades earlier. The mission commander glanced at his pilot. Dropouts in Aster's transmission had left several blanks in the numeric string.

"Some correlate, but no cigar," Mendoza said quietly.

Suspicious by nature anyway, Booth clipped, "Sir, we copy the code, but it's not an exact match. We're gettin' signal dropouts, so . . . Hey! Why aren't you transmitting through military satcom link? Frankly, sir, that's the only way we're *supposed* to get a recall code. We received our *launch* code through Milstar a few hours ago, so tell me: *Why* are you calling us on a commercial satphone . . . *sir?*"

Holding the communicator in his right hand, suspended above the aircraft's center console, the muscular pilot smiled at Mendoza and mouthed, "Spoofing?" It wouldn't be the first time American flight crews had been "spoofed," tricked into thinking a message was from a bona fide commander, when, in fact, it was a very clever ruse employed by an enemy. Happened all the time, during "Red Flag" exercises at Nellis AFB. Mendoza shrugged.

Booth's terse, sarcasm-laced questions, bounced half-way around the globe via multiple Excalibur satellites, arrived in Omaha as broken, barely decipherable snippets, prompting an Aster outburst. "Damn it, Major! If you bomb that ground-based laser, you could trigger a nuclear war! For God's sake! Turn that airplane around—*now*!" Aster bellowed, face reddening.

"General, only a few seconds 'til that Excalibur loses 'em!" Joaquin warned, his voice rising a half-octave.

"Sir! Let me talk to him," Burns demanded, hand outstretched. Aster jabbed the device at his exec in frustration.

Burns tapped his right ear bud and unleashed a loud, machine gun-like litany. "Boomer, this is Burner Burns, your old ops officer. I work for General Aster, and I placed this call to you. Boomer, you get your big Swedish ass out of China *now*, or I will personally strip that tattoo of a vampire off your hairy right butt-cheek! Is *that* authentication enough for you, Major?" Burns, too, was shouting.

In the B-2 cockpit, Boomer Booth frowned. He'd only detected a few garbled words from the new speaker, who sounded vaguely familiar. Most were unintelligible, though, disintegrating in a rush of static.

"What the hell . . . ? Who was *that*?" he muttered. Mendoza was shaking his head. He, too, had failed to understand the new voice's intermittent, broken message. The two pilots stared at each other, each wondering. Turn around and head for home, or press on?

In Omaha, Burns didn't wait for a reply from Booth, knowing the fragile link was breaking. "You shitbirds stay in max-LO mode and head for your tanker rendezvous point out over the water! Move it, *now*! And we had to call by communicator, because the Chinese PLA bastards are jamming . . ." His voice trailed off as he heard increasingly loud static filling the earpiece. Attempts to reestablish a connection failed.

"He's well outa the window, guys," Joaquin finally said, staring at Aster's computer screen. "I think I have a pretty good idea where that big boy is, though. I marked a point when we made contact, and again when we lost him." He pointed to a couple of white "Xs" on the graphic.

"Holy shit!" Aster swore, leaning over Joaquin's shoulder. "He's less than an hour from that laser site! Burner, how far out do you estimate?"

Burns squinted, noting the distance scale Joaquin had highlighted with a cursor. "At high-altitude, in LO mode, staying as stealthy as possible—I'd say they were, uhh . . . thirty minutes from the IP, max."

Joaquin glanced up, puzzled. "LO? IP?"

Burns kept staring at the screen, recalculating time and distance, before responding. "IP's the 'initial point,' where a guy starts his bomb-delivery run. LO means 'low-observable,' basically staying in stealth configuration. The B-2 can retract its antennas and limit control surface travel to make sure it stays as radar-stealthy as possible."

In the darkened cockpit of a B-2 well inside Chinese airspace, Booth finally thumbed a button, terminating the call after several seconds of uninterrupted static. He checked to ensure the communicator was still powered-on before clipping it back onto the edge of his flight suit's pocket. "Now what, Ringo? Press on, or abort?"

The two pilots stared at each other for a long moment. Then Mendoza made the most difficult decision of his stellar career.

STRATCOM HEADQUARTERS/COMMANDER'S OFFICE

Aster absently watched Joaquin virtually twist and turn the graphic of Earth and Excalibur-satellite tracks, searching for another space-craft that might soon have the B-2 bomber in its coverage beam. So far, nothing looked promising for another shot at reaching the bomber. Still, the general mused, the pilots had obviously received his order to abort. Unfortunately, there was no way to track the B-2, no space-based eye-in-the-sky radar to assure him the bomber had aborted its attack on that GBL site near Beijing. Again, Congress had decided spending money on even a rudimentary space-based radar wasn't "cost effective." *Dammit!*

The STRATCOM chief was suddenly aware of how incredibly tired he was. Successive crises over months had consumed a tremendous amount of physical and emotional energy. *Now we're into another one. Those damned hard-headed majors on Spirit 11 might still be pressing toward the Chinese capital city, determined to drop a load of 2,000-lb. precision-guided bombs. Or, maybe they aborted. Maybe the GBL site is safe. Maybe it isn't.*

If Lee's diplomatic efforts succeeded in keeping China's more-militant elements under control, and *if* those two B-2 pilots had aborted and were now headed for the coast, the world just *might* side-step catastrophe. Of course, Chinese missiles could still be launched as a first-strike, rather than in retaliation for an attack on the GBLs. *We're not out of the woods yet.* The general sighed, listening to Hurl-burt's handheld communicator ringing.

Waiting, Aster scribbled a note on a blue notepad topped by four stars. He ripped the sheet off and handed it to Burns, saying, "Shoot this priority e-mail to Admiral Lee! *Now!*" Burns *yes-sirred* and ran to the office door. *God, please make sure Lee gets this in time!*

GREAT HALL/BEIJING, CHINA

Lee touched General Luan's shoulder, drawing him aside. The two seasoned warriors conversed in hushed tones. Mi Soo, the only person who had noticed, watched Luan's expression register disbelief, then something akin to awe or admiration. Together, the men returned to the group, where Lee excused himself for interrupting and turned to Boyer's larger-than-life image.

"Mr. President, the time for diplomatic maneuvering has passed," Lee declared, then looked to Yi. "Mr. Chairman, we remain confident the B-2 mission can be recalled, and I assure you this is *not* the opening move of a war campaign against China. To underscore that assurance, I have chosen a path to prove my country's good faith. I will stand alone at the ground-based laser site, the American B-2 bomber's target. General Luan has kindly agreed to take me there.

"It is my fervent desire that you, Mr. President, and you, Chairman Yi, will accept this gesture of goodwill for the benefit of both nations," Lee said. "Perhaps it will foster a trust our peoples so desperately want you, their leaders, to share. I understand the B-2 may complete its mission, yet this old warrior will go, with honor. I only implore both of *you* to cease all offensive military actions."

Lee answered Yi's shocked expression with one of determination. Mi Soo was confused, her frantic eyes flicking from Lee to the chairman to Luan. Boyer cupped a hand over his speakerphone, then barked at T. J. Hurlburt.

Mi Soo stepped forward and touched Yi's arm, showing deference by dipping her chin and slightly bending at the knees. Her eyes no longer fought tears; they flowed freely. "Chairman Yi, my father often advised you on many critical matters," she said, her voice quivering. "Our divisions of late have come at a most unfortunate time, preventing his wise counsel now. But, I know my father would trust Admiral Lee, and I urge you to do the same. Please, don't let this moment be *our* failure! Nu, *he* is the traitor! You know that. Only Nu's hard liners can destroy us, if we submit to their treachery!"

Luan returned to the now-awkward circle. All eyes were on the young Zhang woman. "Chairman Yi," he said, "the Admiral and I must go now. If you and the American President come to an agreement, my assistant will call and relay your decision."

"General, why should you bend to the American's wish?" demanded the chairman, his expression hard.

"I trust Admiral Lee's commitment to China and to peace. I believe his President, as well. I reject the war that looms, a war with shallow roots in both Chinese and American treachery and miscalculation. I will honor the Admiral's courage and commitment by escorting him to the B-2's target."

Luan nodded to Lee, and the two departed. Luan handed something to a security guard standing at the exit.

A long silence descended over the center, a quiet darkened by a pall of impending doom. Chairman Yi sensed that the world stood at the abyss of nuclear war. He turned to Boyer's image. The leaders of two powerful nations held each other's stare, probing, searching the other's soul. Boyer finally broke the silence.

"I will act, Mr. Chairman, to ensure the safety of China and the United States. I will *not* take us to war. I ask that you heed General Luan's words. Please, do not order strikes on our carrier battle groups."

Chairman Yi was visibly surprised. *'Our carrier battle groups!'*

Those are our counter-targets, if the B-2 hits the laser site! How do the Americans know our plans?

The Chinese helicopter settled onto a chipped-concrete pad, its rotor blades blasting a thin coat of dust from the pavement. Lee turned to China's supreme military leader and offered his hand in thanks. Luan looked away and opened a sliding door on his side, dropped to the ground, then stuck his head back inside.

"Admiral! Come on!" the general shouted. "Did you really think I would allow *you* to steal the day's honor? We will share this glory!" Laughing, he bent at the waist to avoid the slowly rotating rotor blades and jogged toward a knot of soldiers waiting in a barebulb lamp's pool of light.

GROUND-BASED LASER SITE/BEIJING CAPITAL DEFENSE REGION

Admiral Stanton Lee and General Luan watched the sky. They stood together at the entrance to China's primary ground-based laser operations center, the B-2 bomber's target. Both were acutely aware this was ground zero. Even now, the B-2's target-designator symbol could be locked onto the building. *If* the stealthy, batwinged bomber was still inbound.

Luan offered Lee a cigarette, which the older man politely refused.

"Old Russian joke," Luan began. "Two men lined up against a wall. Blindfolded. The Czar's firing squad locks and loads. An officer asks the men, 'Want a cigarette?' One of the men says, 'No, I don't smoke.' The other whispers, 'Max, at a time like this, don't make trouble!'" Lee broke into laughter.

A dozen yards away, a Chinese security detail glanced at the general and his crazy American companion, then nervously scanned the dark, star-studded sky. Lee and Luan were laughing, ignoring the younger troops. The older men knew they'd never see or hear the black-skinned bomber. They might hear a brief, sharp whistle of precision-guided 2,000-pound bombs seconds before the weapons exploded. But the human mind couldn't process such imminent danger fast enough to act, to run, to flatten one's body against the ground.

"Do you think your people will be able to stop the bomber, Admiral?"

"I do, General. Never been one to take uncalculated risks," Lee replied stiffly. "And I appreciate you speaking the truth back there. My apologies for doubting you. Before we departed the center, I received a message confirming that China was *not* jamming our Milstar satellite. The North Korean nuclear blast has charged the ionosphere over a large portion of southeastern China. From geosynchronous orbit, Milstar's weak radio signals cannot penetrate that ionosphere. However, our new Arthur and Lancelot spacecraft swing close to the Earth, ensuring their signals get through. Rest assured, the abort message *will* reach that B-2."

"Ah, such confidence. But then, you were a fighter pilot." Luan smiled, flicking ashes from the cigarette's glowing tip. "One who flies from a carrier takes incalculable risks every day, no?"

"No. I always liked the odds I faced in the air. It is *you* who have taken the greater risks, General, during those years of humping heavy packs in dark corners of Asia and beyond. The mud, the jungles, the frozen tundra—and people shooting at you. People you can actually *see*. No, you faced true danger. A pilot's risks pale by comparison."

Luan smiled, drawing deeply on the cigarette, its red glow illuminating a deeply lined face. His mental calculations suggested the B-2 strike could occur within minutes.

"Are you afraid, Admiral?"

"Why should I be?"

"I fear only that we might have failed," Luan said softly.

"We didn't fail," Lee replied, equally softly. "You came with me, because you, too, had faith in the truth. Others will now follow, my friend."

Luan flicked his cigarette into the darkness and raised his face to the sky. The two warriors stood in silence for long minutes, letting the night embrace them. The communicator in Lee's suit-coat pocket sounded off, prompting the men to look at each other. They slowly smiled, understanding. War had been averted.

As Lee reached for the buzzing device, a sudden flush of pigeons caused him to freeze. He, too, heard the distinctive sound, a noise that had struck fear in even the most hardened warriors huddled in trenches and foxholes since the dawn of air combat: the unmistakable whistle of an incoming bomb.

'19   The Precipice

STRATCOM HEADQUARTERS/COMMAND POST

"Sir, Space Command reports two SBIRS positives from the watch-site near Beijing! Signatures are consistent with two E-JDAM deto-nations approximately one second apart."

Aw shit! What the hell have we done? Lieutenant General Dave Forester's heart dropped. A pair of 2,000-pound Enhanced-Joint Direct Attack Munitions had obliterated the Chinese ground-based laser site. The most-advanced iteration of the U.S. military's extremely accurate GPS-guided bomb, recently "enhanced" by a covert laser-guidance capability, an E-JDAM rarely missed its target.

Outwardly, though, the STRATCOM operations chief remained stoic as his brain assessed the implications of the report. A Space-Based Infrared System satellite had detected two explosions near Beijing. The timing of those blasts could hardly be coincidental. That stubborn B-2 bomber crew had *not* aborted its mission. The highly trained air-warriors had pressed their attack, believing Aster's commercial satphone-delivered recall message was a clever "spoof," something every combat crew had been warned to expect in this day of sophisticated "information warfare."

Forester easily could visualize that Chinese ground-based laser site, now reduced to a smoking crater lined with twisted metal and mangled bodies. Heavy casualties among the GBL crew were a certainty.

"Understand," Forester finally answered dispassionately. "Tell Space Command and NORAD to alert us if they even *suspect* a retaliatory missile launch from China." The B-2 strike would almost certainly trigger a Chinese response, and the U.S. had better be ready. The three-star general mentally retraced the most-likely scenario STRATCOM wargamers had developed hours earlier:

- *GBL attack prompts Chinese retaliatory strike.*
- *Surface-to-surface missiles most-likely near-term response vehicle.*

The wargame-postulated target list ranged from Taiwan and Hawaii to U.S. forces in the region—forces such as airfields in Okinawa and those two Navy carrier groups off the coast of China.

"Sir, Army missile defense forces are on high alert and standing by," the officer reported. "NORAD says missile-launch detection systems are degraded, but considered mission-capable."

The STRATCOM battle cab's quiet, intense crew was fielding a torrent of reports from across the globe. The flood of information was automatically fused and presented as stark, concise "actionable knowledge" on workstation screens in near-real-time.

"The Area Defense Commander at Pacific Command also has assumed ops control of one ABL platform and one in-orbit prototype KKV. Both are in position and ready to assist, as necessary," he concluded.

A flurry of activity drew Forester's attention, as Aster and a small contingent burst through the battle cab's door. The room was getting crowded, filled with grim-faced officers and senior enlisted operators.

"Sitrep, Dave," Aster clipped. "I know the GBL site's been hit. Admiral Lee was *at* that site, you know." Forester's stunned glance was answered by the four-star's silent stare. The combat-honed operations director comprehended, all right. But the shocking loss of that stately retired naval officer hit Forester with the force of a body slam. And it underscored the seriousness of an escalating situation.

"Sir, Army One-Hundredth-Missile Brigade crews are on full alert, standing by for shoot orders," Forester croaked, his voice cracking. He cleared his throat, trying to dissipate a lump. "The Airborne Laser platform's also on-station off the west coast of the Korean peninsula. It's orbiting at 38,000 feet, primed and cocked to engage ballistic missiles, during their boost phase. The Army's missile-defense element has maneuvered our one-and-only prototype Kinetic-Kill Vehicle into a transitory position, too. It's now in a LEO orbit, where it *might* be able to engage anything coming from south-eastern China or North Korea—but only in the next hour or so."

"Damn! Those are both unproven missile-defense systems!" Aster declared, fighting to maintain composure. "The ABL's knocked down a few missiles during controlled tests, but the KKV hasn't done a single intercept—except in simulations. It's only been in orbit a few months! I hate to depend on *them*!"

Forester shook his head. "They're insurance, sir. Based on your heads-up, plus wargame projections, the missile defense guys scrambled everything they had and cobbled together a layered, defense-in-depth capability. It's damned ad hoc, but it's all we have."

He turned to a workstation monitored by an intense Air Force major, the same officer who had explained the ramifications of North Korea's nuclear detonation in the upper atmosphere, Aster noticed. To the STRATCOM four-star, that Korean blast seemed to have been years ago, unleashing a rapid-fire series of global crises.

"Major, how 'bout a rundown of the current missile-defense posture—and its limitations," Forester snapped. The officer quickly outlined the status of the Airborne Laser and kinetic-kill vehicle, as well as Army-controlled ground-based interceptor missiles in Alaska and at Vandenberg AFB, California. The U.S. was as ready as it could be to counter any missile threat from China or North Korea.

"The Army interceptors can only be employed if a ballistic missile is inbound to the U.S. or Canada, though. They can't engage threats to Taiwan, South Korea, or Japan, sir. Too far away," the officer clipped.

Aster was silent, studying color-coded graphics displayed on the command center's large screens, visible to everybody in the room. Dotted-line red wedges showed the projected tracks of missiles that could be launched from known sites in North Korea and southeastern China. Translucent-white, fan-like patterns depicted zones where U.S. missile defense systems could engage those threats. He was struck by how fragile and one-dimensional the nation's Pacific Region missile-defense network remained, after almost two decades of development and testing that had consumed billions of dollars.

Missile defense is still a bitch of a challenge, technically and *politically*, Aster concluded. He only hoped America's thin missile shield wouldn't be severely tested over the next few hours. He didn't like the odds.

Aster and Forester conversed quietly for a few minutes, then the STRATCOM commander and his small entourage departed. The battle staff was left to orchestrate powerful U.S. forces facing their most-demanding strategic crisis since the Cold War.

I hope we're up to this, Forester reflected. He took a sip of scalding coffee, knowing the brew would only exacerbate the knot in his gut.

GREAT HALL/CHAIRMAN'S CONFERENCE CENTER/BEIJING

As information about casualties and damage at the ground-based laser site near Beijing trickled in, chaos reigned in the Great Hall conference center. Mi Soo, lost in the swirl of activity, was consumed with fear and worry for Admiral Lee, General Luan, and the people of two great nations. She also felt strangely removed, viewing the room's occupants from afar. They seemed to move in slow motion.

A furor near the entrance caught her attention. Dozens of senior PLA officers and Party leaders were streaming into the center. Several appeared stunned and disoriented; others were furious. In general, the group was fiercely indignant and irate. A defiant clamor for retaliation dominated.

Someone bumped the now-abandoned video camera that had

linked Chairman Yi to President Boyer, unintentionally aiming the lens at the center's large Knowledge Wall. It now beamed only a cocked-angle image of the wall's large displays to Washington.

Over the din, Mi Soo could barely hear the American president pleading for Yi to respond, his distressed image still hovering above the center's majestic table. However, the chairman's attention was diverted by the new commotion. Surrounded by shouting, arm-waving men, he was no longer listening to his U.S. counterpart. One of those agitators slammed a hip against the conference table, causing Stanton Lee's abandoned computer to awaken. Its display's sudden illumination drew Mi Soo's attention. Two curt, brief text messages on the screen triggered a muted cry of anguish.

Lee: Unable to recall Spirit 11. High probability of B-2 strike on GBL site within the hour. Recommend immediate withdrawal.—Aster

Mi Soo had not seen the former admiral keying an answer, but there it was, Admiral Lee's reply: *Aster—Copy all. Thanks.—Lee*

A tiny-font third message near the screen's lower edge brought tears to the young woman's eyes, making it difficult to read:

Mi Soo: You are a remarkable patriot. Carry on; do what is right. We will meet again.—Lee.

Tears spilling, she reread the messages, then erased the personal one to her. She ripped a connector free, scooped up the small computer and dove into the crowd, trying to elbow her way through the cluster of bodies huddled around the chairman. Angry men were demanding Yi order an immediate retaliation. Cradling the computer, the diminutive woman shoved an arm through the jostling mass, trying to draw Yi's attention. A red-faced PLA general angrily slapped her arm away and elbowed her aside.

Circling the group, Mi Soo frantically tried again. "Mr. Chairman! *Please!* You must see this!" she demanded, her high-pitched voice competing with the cacophony.

The same rude general forcefully grabbed her arm from behind, spinning Mi Soo around. With the other, she juggled the open-screen notebook computer, fighting to hang onto it. The officer's dark face was thrust into hers, his mouth twisted. He stabbed a finger at the wall's big screens.

"*Look!*" he shouted. "Do you not understand, woman? American bombs *destroyed* the laser battery and *killed* General Luan! Many soldiers are dead! Brave Chinese brothers killed by imperialist aggression! Their deaths *will* be avenged! *Go away!*" he growled harshly, shoving her.

Teeth clenched, Mi Soo plunged into the crowd, straining to hear Yi over the noise. He was ordering a launch against the American carrier battle groups.

"Mr. Chairman! *No! Oh, my God, no!*" she cried in English, inadvertently using an American idiom of alarm. The incongruous plea worked. Yi turned, eyes searching. Among the broad uniformed backs and shoulders, Mi Soo glimpsed the chairman ordering men out of the way.

"You! Zhang girl! Come here!" he ordered sharply. The crowd parted, offering an opening. She plowed through to the chairman's side, holding Lee's computer in front of the man and speaking rapidly. Yi tipped his chin up to read the screen's two messages.

"Admiral Lee *knew* the American bomber would hit the GBL battery!" she cried. "The attack was a terrible mistake, and he knew it couldn't be prevented! Yet, he sacrificed . . ." Her words were drowned out by another loud bark.

"Mr. Chairman! General Luan *also* knew the American bomber would strike! You must honor this hero of the revolution by avenging *his* sacrifice! To hell with the American dog!" a tall Chinese officer demanded loudly, shoving a small, handwritten note to the chairman. Expressionless, Yi glanced at the man and accepted the slip. Mi Soo hovered close to the chairman, fearing she'd be pushed aside again.

He scanned the note, then read it aloud, his words strained: "General Luan says, 'Mr. Chairman, I implore you to reinforce the order I have issued to Fleet Command: *Halt all hostile actions.* Do *not* order retaliatory strikes against U.S. forces, regardless of the American bomber's success. I go with Admiral Lee, on your behalf, to show America that China is committed to ending this crisis peacefully.—General Luan' "

Chairman Yi locked eyes with Mi Soo, sadness and pain reflected in his. He slowly shook his head, saying, "I cannot honor the general's request, little one. I am sorry . . ." Yi turned away, speaking rapidly. Several officers smiled and ran to the command consoles, issuing orders.

Resigned, Mi Soo slumped. In a span of minutes, events had taken a terrible turn. And the consequences were only too apparent.

Oh, dear father! she thought, fighting new tears. *We have failed! China and America are at war!* Tears flowed freely, tears for her failures, for her country, and for Admiral Stanton Lee.

OFFUTT AFB/STRATCOM WARGAMING CENTER

General Howard Aster burst into STRATCOM's wargaming center and joined Colonel Androsin in the pit. The commander was still talking to his handheld communicator, nodding vigorously and responding curtly, "Yes, sir . . . Understand, sir."

Covering the device's microphone, Aster tipped his chin toward Androsin, stage-whispering, "Jim! Any damage assessment yet?"

Androsin pointed to his monitor. "Yessir. Preliminary data from SBIRS indicate the GBL site was leveled. No comm from Admiral Lee . . ."

Aster raised a palm. "Yes, Mr. Secretary. The ground-based laser site *was* hit. Damage assumed to be extensive. No word from the B-2 crew . . . Yes, we believe Admiral Lee was at the site . . . Yes, sir. My assessment is: We've committed an act of war against China, even though it was unintentional. Recommend you advise the president that STRATCOM is postured to exercise flexible deterrent options."

After a pause, Androsin noted the general's jaws were flexing, a sure sign of impatience. "Okay, we'll bring up the center on our battle screen." Aster jabbed a finger toward a large wall display and clipped a quick, whispered order, prompting Androsin's hands to fly across the wargaming pit's control panel. An image of the Chinese conference center's Knowledge Wall snapped into focus, as Aster walked closer to the big screen. The image from China was askew, but clear enough to read data on the center's displays.

"Yes, sir. I can read it . . . Definitely looks that way . . . Okay, I'll see if they're willing to help. Assuming they are, you should see the text change in a few minutes." Aster snapped his communicator closed, ending the call, then turned to Androsin.

"Jim, how 'bout rounding up two of our media types in the wargame-coverage pool—Dane Winfield, the *Aviation Week* guy, and whoever's reporting for the *Kosmos* GloCon site. Get 'em both in here ASAP."

Minutes later, Aster outlined the rapidly deteriorating situation for Winfield and a *Kosmos* GloCon reporter, pointing at the wargame center's large primary screen. Colonel Marie Avery, the command's public affairs chief, offered a few suggestions to clarify Aster's request, then bluntly asked, "Well, guys, are you up to helping us avoid World War III?" Both men murmured an affirmative, stealing glances at a one-page summary Avery had given them. "Okay. Put your own spin on it, if you want. Just get something up on your Web sites ASAP. We're running out of time."

Aster called Hurlburt and informed him the two reporters had agreed to jump into the fray. Keeping Aster on the line, Hurlburt relayed the hastily concocted plan to the president, who was trying to reengage Chairman Yi via videoconference. So far, the Chinese leader was pointedly ignoring him. The videoconference camera in Beijing was still aimed at the center's Knowledge Wall.

Aster and Androsin stared at the big screen in the STRATCOM wargaming center, waiting. Androsin suddenly pointed, prompting Aster to report via communicator, "Yes, we see it, sir."

For the first time in days, it seemed, Aster felt a glimmer of hope and a lopsided grin spread across his tense features. On the Chinese Knowledge Wall's third tactical display, text started rolling, drawing the attention of operators, who started pointing and chattering in Mandarin.

Mi Soo, slumped in the chair Stanton Lee had occupied only an hour earlier, reacted to shouted alerts. Operators were pointing at the huge third screen overshadowing the center. She wiped her eyes, trying to clear the blur. Reading the scrolling text, she, too, started smiling, instantly grasping what the Americans were doing. First, the *Kosmos* GloCon Web page, then the adjoining *Aviation Week* text, described in graphic, stark terms the STRATCOM wargame projections emerging from Lee's "China side game."

"If the B-2 were to strike China's GBL site," the GloCon reporter was writing, *"the Chinese PLA will retaliate and strike specific target sets."* He listed American cities, airfields, and two carrier groups in the Northern Pacific that China would probably strike within hours.

The projections elicited smug smiles and confident glances among the center's PLA officers, Mi Soo noted with alarm. Soon, though, the smiles vanished, the center grew quiet and grim faces proliferated. New GloCon and *Aviation Week* text flowed across the screen, seemingly in real-time, as writers in Omaha typed directly onto their Web sites.

Chairman Yi was the first to whisper an exclamation, loud enough to prompt sideways glances from aides and uniformed PLA officers clustered around the nation's leader. Mi Soo silently coaxed the stark messages from that large screen, willing two writers half a world away to convince her countrymen.

Both Web sites' rapidly scrolling words were describing equally dark descriptions of probable U.S. responses to China's retaliatory attacks. Shanghai, Beijing, and other cities would be targeted, possibly with nuclear weapons, the wargamers had determined. In short, China's political and military leaders were reading what they knew instinctively, but had never internalized. However, seeing high-probability predictions flow from America, written in graphic terms that could easily be construed as the history of the future, those leaders were forced to confront a lie they had come to believe: America would never, ever "go nuclear," during a U.S.-China dispute.

Mi Soo had seen the Knowledge Center's wargame outcomes, and had argued vehemently against China's false premises, which had led to such false conclusions. She had studied the unclassified American wargame outcomes, and had seen the intelligence stolen by China's legion of spies in the U.S. and Europe. She knew America definitely *would* go nuclear, under the right circumstances. And the current tit-for-tat exchange of deadly strikes was precisely the scenario that would escalate to both nations firing nuclear warheads across the Pacific.

Off-camera in Washington, President Boyer had snatched the handheld communicator from his defense secretary, T. J. Hurlburt.

"General!" Boyer barked to Aster. "Are you seeing the same thing I'm looking at in China?"

"I am, Mr. President. Let's hope it changes some minds over . . ."

Boyer interrupted brusquely, "I want your guarantee that you or any other combatant commanders have *not* ordered *any* global strike missions against China! You *will* hold your forces in place! Are we absolutely clear on that?"

"Loud and clear, Mr. President. But it's my job to advise you in times of strategic crises," Aster fired back, stubborness born of long days and tension struggling to explode. "Well, we're damn sure there, sir. *Now!* We *are* at war! If we don't take immediate defensive measures, we risk absorbing multiple, potentially devastating blows. American lives are at risk, sir! A hell of a lot of lives! Unless we act soon, our capability to *respond* will be in jeopardy, as well."

Ignoring Aster's plea, Boyer saw the Chinese leader reappear on the videoconference screen, and exclaimed, "Chairman Yi! *Can you hear me?*"

Unasked, Androsin adjusted the Pit's controls to place side-by-side images of Boyer and Yi on the wargame center's primary display. Glancing sideways, he also noted STRATCOM'S four-star general's jaw muscles twitching again, his face set in hard, deep lines. The commander was more pissed than the Army colonel had seen him in ages. A stir of whispers rippled across the hushed wargaming center, as the U.S. president addressed China's supreme leader.

"Mr. Chairman," Boyer began, his tone firm, yet calm. "As you know, we were unable to recall the B-2 bomber in time to prevent this terrible, terrible mistake. I assure you, however, that no other offensive actions are directed at the People's Republic of China. *None!*"

Aster noted a similar stillness had settled over the Chinese conference-and-command center. Every person there, just as those in STRATCOM's center, was acutely aware that each passing second was a tick closer to war, an all-out, no-holds-barred conflagration now being described in too-graphic terms by Dane Winfield and the *Kosmos* GloCon writer. Faced with what, in essence, were the unflinching facts of both nations' warplans, each point countered by a more deadly counterpoint, human beings on two continents were forced to pause, to think—and, for the moment, to listen. On opposite sides of the world, dozens held their collective breath. Only those two people on-screen had the power to stop an imminent nightmare of carnage.

Yi wasn't looking at the videoconference camera, seemingly ignoring Boyer's passionate plea. He was reading text still scrolling on the Knowledge Wall's third screen. After what seemed an eternity, Yi stepped forward and stared into the videoconference camera.

"I am certain your Admiral Lee died with the conviction that peace was at hand, that we had avoided war. But America—you, President Boyer—have chosen a different path. I must protect China's people and its sovereignty, just as you would in my position. I have already ordered appropriate actions to deal with immediate threats. And I am prepared to take this fight to your shores, your cities, if necessary."

"Holy shit!" Aster muttered.

Boyer tried to object, but was cut off by a dismissive, curt reply from China. With that, Yi turned and walked away from the camera. Aster, standing next to Androsin, whispered, "That son-of-a-bitch! He snubbed the President of the United States! He declared war!" Equally stunned, Androsin could only nod, eyes locked on the wall-size screen.

Boyer turned to T. J. Hurlburt, off-screen. With no apparent emotion or attempt to conceal his words, the president said, "T.J., order General Aster to execute Plan Aurora." Off-camera in Beijing, Yi's primary aide had heard Boyer's seemingly private order, and dutifully relayed it to China's chairman.

In the STRATCOM wargame center, Aster's communicator sounded off. The Secretary of Defense ordered, "Execute Plan Aurora." Aster relayed it to Dave Forester, via handheld communicator.

Grabbing a wireless keyboard, Androsin retrieved the command's so-called "Initial Warning Order," then directed Aster to its presentation on the wargaming Pit's private display. The general quickly scanned the "Aurora" outline, refreshing his memory.

Upon issue of an order, a series of events would redirect America's Single Integrated Operational Plan to execute Plan Aurora.

That would initiate a flurry of intentionally confusing, secure communications with the national command centers of Britain, Russia, and India. Aurora also called for stepped-up communications traffic with the U.S. Embassy in Hanoi. But these were designed only to *appear* as such. Reality was something else entirely.

A sophisticated psychological warfare campaign, Plan Aurora was a relatively new product of STRATCOM's wargaming team, a clever deception cloaked in just enough security to deny most counter-intelligence probes. But, like a miniskirt, it also provided enough visibility to attract concerted attention. A sophisticated intelligence service, with little effort, would uncover the plan's general premise.

Aurora's flurry of U.S.-initiated communications with national command centers was a sophisticated netcentric-warfare ruse launched by STRATCOM's powerful BOYD computer. Carried out by a secret Strategic Command detachment, Plan Aurora specifically targeted China. It was designed to create doubt and confusion, prompting a strategic pause in any large-scale Chinese military actions the nation's leaders might initiate. Aster hoped to heaven it would succeed.

Jill Bock suddenly appeared at Androsin's elbow. "Move and countermove," she stage-whispered. She taped a grainy, handheld communicator-captured photo to the pit's workstation. Aster did a double take. Admiral Stanton Lee's frozen, unsmiling image stared at him, bushy white eyebrows squeezing deep frown lines between them. A thin black stripe angled diagonally across the photo's upper right-hand corner, a subtle Jill Bock tribute.

Aster turned back to the center's primary screen in time to see President Boyer disappear from half of the wargaming center's big display. The U.S. president could hold his own in an international "up-yours" game of snub-trading, Aster decided.

STRATCOM HEADQUARTERS/COMMAND CENTER

"General! NORAD's reporting multiple launches from China!" a Navy captain shouted, causing Forester to choke on his coffee.

"Southerly tracks! Initial projections indicate no factor for Hawaii! Repeat: Hawaii's *not* the target!"

The carriers! Forester flashed. He punched a code on a handheld communicator, then adjusted its wireless earpiece. Aster and the entire U.S. command chain would be in his right ear within seconds. A dozen faces turned toward him. Every man and woman in the battle cab had frozen in place.

"Activate missile defenses," Forester ordered. "Tell Army missile defense crews to stand by; PACOM's cleared ABL to take the first shots. Hold all other missile defense assets! Until ordered otherwise, only the ABL is cleared to engage! Report intercept outcomes!"

Forester grabbed the Air Force major's shoulder and added, "Lance, alert those carrier groups; they've got incoming missiles. They'll have to handle any leakers the ABL misses, so they'd better be ready to fight!"

"Roger, sir. The Lancelot bird's just coming into line-of-site with the carrier groups."

"General, NORAD's got a classification! At least four Chinese M-19s in the air!" the Navy captain interjected. *M-19. New type,* Forester recalled. *Two-stage, medium-range, liquid-fueled; either nuclear or conventional warheads.*

Forester heard Aster's communicator ring. All hell was about to break loose . . .

YAL-1A AIRBORNE LASER AIRCRAFT/38,000 FT. OVER PACIFIC

What the bloody devil am I doing here? Damon "Deke" Creede grumbled to himself. He focused on a flat-panel display blanketed with symbols and graphics showing the status of his pride and joy, a massive chemical oxygen-iodine laser, or COIL. The huge, complex laser weapon dominated the cavernous cargo compartment of the YAL-1A, a converted Boeing 747F freighter. Creede was on that aircraft, strapped into a crew seat at the COIL's fire control console.

Civil-servant weapons test engineers aren't supposed to be in a damned combat zone! he screamed to himself. The career federal-government engineer tried to convince himself he wasn't scared, but a chronic urge to pee said otherwise.

He had deployed to a U.S.-occupied air base in Japan only a day earlier, part of a hastily assembled crew drawn from the Airborne Laser test force at Edwards AFB, California. At first, Creede had been honored and excited when Lieutenant Colonel Mark "Rabbit" Kleiner had tasked him to serve as the ABL's primary laser "shooter," a role he'd filled, during brief, but intense, preliminary operational-suitability tests of the YAL-1 and COIL.

But excitement had turned to abject dread when he'd learned *why* the Air Force's one and only Airborne Laser platform was be-ing yanked from its test program and sent to the Pacific, even though it had not been declared operational. Increasing tensions in the region had raised Washington's concern that North Korea or China might do something stupid—like launch a missile against an American target. The North's nuclear weapon detonation in the upper atmosphere a few weeks ago was responsible for the ABL be-ing scrambled, prepared to take on any other missiles the crazy bas-tards might launch.

But Deke Creede was hardly combat material. By his own admis-sion, he was the stereotypical civil servant, a long-time weapons test engineer who'd built a modestly successful career at the Air Force's hot, windy, sprawling test facility in the Mojave Desert. Overweight and balding, he'd never worn a military uniform and was the an-tithesis of a warrior. He was simply a classic geek-engineer who hap-pened to be fascinated by big lasers and had enough seniority to wrangle an assignment to the ABL project. If the frickin' Congress didn't kill it, a years-long ABL test program should get him to career-completion and a cushy retirement check. Then *this* crap had hit the fan!

"Hey, Deke! Go button four!" an excited, young enlisted commu-nications operator, also seated in the modified Boeing 747's sealed battle module, called via interphone. "Got an alert from STRAT-COM through the Army missile defense battalion!" Deke selected a menu on his comm panel, scrolled to the designated frequency and punched a small square switch on the multifunction display.

". . . tracks are southerly; no indication of maneuvering; standard ballistic trajectory. Dragon, you're cleared to engage. Repeat, *cleared to engage!*" he heard.

"Dragon copies, and will engage," Kleiner responded via satcom radio. "Deke, you copy that?" The left-seat pilot and ABL test force commander's familiar voice came from the 747's flight deck, this time over the aircraft interphone.

"Some of it, Cap'n. Is this shit for real?" Creede's brain hadn't grasped the full impact of that partial, level-toned missile-defense crew transmission.

"As real as it gets. This is what our toy was designed to do. Make it work, Deke," Kleiner said, calm, but with an edge.

Creede gulped and glanced to his right, where a young captain was already setting up an acquisition and tracking solution. A separate low-power laser, nestled in a pod mounted on top of the 747's prominent hump that graced the stately aircraft's forward fuselage, would lock onto an incoming missile, track it, and provide a firing solution for the massive, high-powered COIL.

"Okay, Deke. I've got three, maybe four ballistic tracks. They match SBIRS cues. I'm engaging," the captain said. Creede shook his head, still trying to accept that this was the real deal, not a test. *Get your act together, Deke! Two, maybe three shots; that's all you'll get. Make 'em count . . .*

On Creede's battle display, a white diamond surrounded the tip of an arcing red track-line. The Active Laser Ranger acquisition-and-tracking system, a modified version of the Lantirn targeting pod used so effectively by fighter/attack aircraft, was locked on to one of the Chinese ballistic missiles. The diamond shimmered as it moved slowly skyward on Creede's display, then converted to a dancing green box. The pod's laser had fired, its energy bouncing off the missile's body in space, giving the YAL-1A crew an accurate range-and-bearing track. Deke's COIL had a target.

Breathing deeply to steady his racing heart, Creede willed professionalism to take control. Just as he had done in countless simulated exercises, as well as half a dozen actual COIL tests, he coolly assigned the designated track to the COIL's fire control system. On the aircraft's nose, a gargantuan, gimbaled turret twisted and turned, then locked, its glass eye staring up and to the right.

That unblinking eye was cued by the "acq-and-track" laser to find a long, thin rocket body standing on a tail of white-hot fury. Deke had no idea what target it was seeking, but the missile, and those in trail behind it, had been declared a threat.

Show time—Locked and tracking. Laser's powered. Creede had the checklist memorized, but still glanced at a plastic-covered printout, not trusting his excited mind to remember each critical step. A screwup meant the megawatt-class laser might not fire, or might miss its target, and a miss meant someone would die. *Not on my watch, you bitch!* he declared.

"Shooter's ready to fire," he announced, thumbing a hinged, red shroud to expose a rectangular, now-flashing lighted switch.

"Clear to fire," Kleiner responded. The aircraft cruised through smooth air, wings level. Except for the gimbaled nose-mounted turret, distinctive "acq-and-track" laser pod on the "roof," and the muted-tone bars-and-star USAF symbol on the wings and tail, the 747 could have passed for a gray-skinned airliner carrying a load of passengers across the endless Pacific.

"Stand by—*Fire one!*" Creede called, pressing the switch. Chemicals flowing through hundreds of feet of plumbing in the aircraft's cavernous cargo hold instantly mixed. That stimulated a rapid, yet precisely controlled reaction, creating megawatts of laser light. The light flashed through a coated optics train to a giant eye in the nose turret. A powerful, tightly focused laser beam blasted through the polished-glass window, traveling at 186,000 miles per second, aimed at a tiny spot in space filled by a missile body festooned with a Chinese red star. The YAL-1A's sophisticated adaptive optics automatically modified the beam's structure, compensating for air turbulence and ensuring the laser remained in a tight, cylinder-like ray of deadly energy.

The missile was climbing, still in boost phase, well below maximum altitude, when the laser beam slammed into its ice-cold tubular body. With a million-plus joules of energy searing a spot smaller than a basketball, the missile's aluminum-alloy skin expanded and ruptured, allowing the laser's heat to ignite rocket fuel.

The explosion's flash was spectacular. A bright, expanding source of light and heat was duly recorded by a SBIRS satellite. Within seconds, an alert had flashed through radiation-hardened relay satellites to ground stations, on to NORAD/NORTHCOM's Space Control Center at Peterson AFB, Colorado, then STRATCOM's command post in Nebraska. Sensors on the YAL-1A ABL aircraft had already detected the explosion in space.

"Splash one, Deke! You *nailed* the sucker!" shrieked a sensors specialist. A cheer erupted over the ABL's interphone, but died quickly at the sound of Rabbit Kleiner's sharp command.

"Knock it off! Threats still in the air!" he yelled. "Deke, can you get another one? Now confirmed, four of 'em . . ."

"I'm locked and tracking Number Two," the acq-and-track captain said calmly. "You'll have ranging in a second."

Creede had already recycled the big COIL, simultaneously checking pressures, temperatures, and dozens of other parameters. "We're cool here, Cap'n. Laser's comin' back to stable conditions, power's lookin' good. We're 'go' for a second shot," he snapped. Sweat was running down his temples, but both hands were steady as they skimmed a keyboard. A few trackball tweaks and a couple of clicks on valve icons confirmed his optimism. The big laser was ready to fire again.

"Locked and tracking. There's your ranging! All yours, Deke," the captain called.

"Stand by—*Fire!*" Creede shouted. He couldn't help himself. He was both excited and scared silly. *Don't screw up. Don't screw up. Stay cool . . .*

Again, chemicals blasted together, creating another powerful beam of light that spat from the nose turret, pierced the thin upper atmosphere and streaked into space. Another missile body succumbed to the directed-energy weapon, disintegrating in a space-silenced flash.

Creede unconsciously held his breath, waiting for final confirmation from eyes-in-the-sky satellites. But a flashing icon suddenly said he had bigger problems. "Cap'n, we're in trouble!" Creede warned over the interphone. "Got an overheat in the nose turret. Something's goin' bad up there!"

In the cockpit, Kleiner scanned a special display devoted to the laser system. It was far less detailed than the one Creede and others in the battle module were monitoring, but the message was the same. A flashing red TURRET OVERHEAT message said it all. He made an instant decision, one rehearsed hundreds of times during mock emergencies in the ABL simulator.

"Shut 'er down, Deke. The optics train's too hot." *Damn! Two shots and we're Winchester!* Kleiner was acutely disappointed. Yet, he knew two no-shit combat kills with a developmental ABL were more than anybody in the Pentagon would have dreamed possible for such a complex, immature system. He couldn't take a chance on damaging the huge, expensive COIL, or risking an airplane full of people.

"*Splash two!* Hot damn, Deke!" the young sensors specialist shouted again. "You're almost half an ace, dude!"

Creede couldn't answer, but flashed a weak grin. He was busy shutting down the COIL, assisted by several flight-suit-clad officers in the close confines of the ABL's battle-management module. Sealed, air conditioned, and pressurized to ensure no leaks from chemicals in the back could seep in, the box-like module suddenly seemed much too small and crowded. But they'd done it. They'd shot down two Chinese missiles—in combat! He was vaguely aware that he and his crew had just made history. War at the speed of light had begun, and he—Deke Creede of pastoral Tehachapi, California— had pulled the first laser-weapon trigger.

20 MISSILE DEFENSE

STRATCOM COMMAND CENTER

"Two down. Two still in the air, sir," Forester reported. In the command post's battle cab, Aster was juggling two phones, one to field the latest from his wargamers upstairs, and a second for relaying near-real-time updates to the SecDef, who was standing at President Boyer's elbow. The president, Forester knew, was still linked via videoconference to Chairman Yi in Beijing.

Forester leaned close to another workstation, reading a flash message. He straightened, frowning. "That confirmed, Captain?" he asked.

" 'Fraid so, sir. ABL's out of business. Optics overheated," a Navy captain answered.

Forester turned to Aster. "General, ABL's out of commission. I'm clearing Fourteenth Air Force for a KKV shot." The operations chief barked orders to another officer, who acknowledged and turned to his workstation.

Within seconds, an Army officer in Fourteenth Air Force's Joint Space Operations Center at Vandenberg AFB, California, received an order he never dreamed he'd hear. The JSpOC commander had cleared him to shoot down a Chinese missile arcing through low-Earth space. Thanks to rigorous training, he was ready—and so was the nation's one-and-only experimental Kinetic Kill Vehicle cruising in orbit.

The KKV's sensors, cued by signals relayed from the ground, were steadily tracking two potential targets that had exited the atmosphere at supersonic speeds. Signals from a ground station brought the KKV from standby to a state of combat readiness. About the size of a small-block Chevy engine, the vehicle was a mass of glass-eyed sensors, fuel lines, electronic packages, antennas, and small rocket nozzles. Within seconds, bursts of flames spurted from several of those nozzles, accelerating the compact vehicle to a speed measured in thousands of feet per second.

Onboard computers performing millions of calculations per second continually refined an intercept solution. Sensors tracked the white-hot rocket plumes of both Chinese missiles, but a highly classified onboard radar system's microprocessor made the decision: Attack the lead missile. It was the most immediate threat. Divert rockets fired briefly, refining the KKV's trajectory. A burst of data from an S-band antenna assured JSpOC operators that the KKV was locked on to Threat Alpha.

Hundreds of miles that initially separated the KKV and its target vanished in seconds, blurred by staggering speeds that had come to characterize the merger of space warfare and missile defense. In the final endgame phase, optical sensors took over guidance control, backed by inputs from both infrared and radar systems. Onboard microprocessors received sensor data, compared them, threw out those deemed unreliable and fired commands to the guidance-and-control system. Tiny pulses spitting from divert-rocket nozzles on the vehicle's exterior increased in frequency as the distance closed: *Pow!* *Pow!* *Pow!* Each firing adjusted the KKV's trajectory by tenths, then hundredths of a degree, making sure the interceptor and its target missile would arrive at the same spot in space at exactly the same moment.

The Chinese missile's ghostly white body grew rapidly in the optical sensor's unblinking eye, the image transmitted by telemetry signals to Earth. Because the KKV was designed as a test platform, not an operational weapon, ground-based commanders, engineers, and technical specialists were treated to unique real-time views of a live intercept against a validated missile threat. The unprecedented intercept was by no means guaranteed. When literally trying to hit a bullet with an even faster bullet, anything could go wrong.

The last seconds were a flood of repeated see-assess-react cycles as the KKV closed with its target, then slammed into the missile. The Chinese rocket disintegrated. The shock of impact had instantly migrated to the warhead, which detonated, its conventional, non-nuclear explosive creating yet another flash detected by SBIRS and DSP platforms in geosynchronous orbit. Seconds later, shouts of released tension echoed through the warrens of NORAD/NORTHCOM's and STRATCOM's command posts, as well as the JSpOC at Vandenberg AFB.

STRATCOM HEADQUARTERS/COMMAND POST BATTLE CAB

"They got the son-of-a-bitch, sir!" proclaimed the Navy captain, pressing his earpiece closer to hear a detailed report from NORAD. Forester grinned and raised a clenched fist in triumph, relieved. But there was at least one more missile on the loose out there. He prayed there weren't more being prepared for launch in China.

He clipped new orders to his battle staff, which assured him the Navy was ready. *Last line of defense,* Forester reflected. His gut contracted yet again as a vision of several ships clustered in a bull's eye crossed his mind.

No Star Wars lasers and fancy hit-to-kill space weapons left. We could be down to old-fashioned bullets in a few heartbeats.

U.S. NAVY TICONDEROGA-CLASS DESTROYER/ SOUTH CHINA SEA

The Aegis combat system's SPY-1D phased-array radar had received cues beamed from a SBIRS satellite to Pacific Command headquarters, then back to the ship via that blessed Lancelot spacecraft soaring overhead. But the limited-capability minisatellite wouldn't be of use for long, thanks to its low-Earth elliptical orbit. Soon, its diving, then climbing track would place the vehicle out of range, unable to aid the carrier groups. Whether downlinked cues from PACOM would reach the ships' missile defense systems depended entirely on the vagaries of Lancelot's orbit and luck-of-the-draw timing.

The ship's radar, cued by SBIRS data through PACOM, searched the sky for a single incoming missile on a ballistic trajectory. Sailors in the darkened intercept-control center calmly monitored radar inputs, rechecked the readiness of several Standard Missile-3 (SM-3) interceptors poised in vertical launchers and mentally rehearsed multiple what-if scenarios. This was real. No "aw-shit" errors allowed this time. No post-training-exercise cursing and humble apologies. Either they got it right or someone, maybe them, would die.

"NORAD/NORTHCOM reports one leaker. ABL and a KKV took out three of the M-19s, but a fourth is inbound; altitude approximately 230,000 feet. Trajectory indicates our group's within the target footprint," a chief petty officer announced matter-of-factly.

"Copy all. Stand by to engage," the combat commander said, his calm tone matching the chief's. *How in God's name did the Chinese target a mass of moving ships at sea?* the officer mused, intrigued by the difficulty of such a task. Maybe China had secret, radiation-hardened satellites tracking the two carrier groups. Intel hadn't mentioned Chinese satellites, but intel seemed to miss a lot these days.

"Approaching engagement envelope, sir," the chief reported, assimilating multiple inputs from other operator stations around him.

"Prepare to launch Standard-One," the commander clipped. He resisted an urge to hover over the chief's shoulder, wanting to see for himself the no-shit M-19 threat boring down on their position.

"We're hot, sir. Ready to fire on your command."

"Intercept solution confirmed. Cleared to fire!"

The weapons control officer punched a "FIRE" button and reported, "Standard-One launch!"

On the deck, fire and smoke spewed from open hatches, then a long, telephone-pole-like missile jumped from its nest, climbing on a long spear of blinding-hot flame. The solid-fueled interceptor initially streaked vertically, then lazily nosed over, leaving an arc of thick white smoke to mark its path. A muffled roar echoed inside the command center.

"Stand by to launch Standard-Two," the commander said, ignoring the first missile's fading rumble. The book called for restraint, husbanding the supply of SM-3 interceptor missiles, if an unknown number of threats might be encountered. As far as he knew, there was only one bogey left out there. If it somehow eluded the first Standard, he wanted a second interceptor in the air as a backup.

"Standard-Two's hot. Fire on your command," the chief repeated, eyes flashing between screens. On the first display, he monitored Standard-One's track, watching the gap close between its climbing trajectory and the radar-predicted track of the incoming Chinese M-19.

Silently counting to ensure the second interceptor wouldn't follow its predecessor too closely, the commander finally ordered, "Intercept solution confirmed. Cleared to fire!"

Again, the weapons control officer jabbed a FIRE button, sending a second SM-3 into the air. Its path was marked by a second streak of white smoke that paralleled that of the first missile. As its roar diminished, personnel on the upper deck shielded their eyes from a just-rising Sun, trying to spot two missiles streaking into the stratosphere.

High above the Earth, Standard-One's sensors, cued by the ship's Aegis radar, searched a small square of sky for its quarry. They found it, locked on, and started talking to onboard guidance-control circuits. The microchips calculated tiny refinements, commanding control fins to twitch several times a second, ensuring the Chinese target was in Standard-One's crosshairs. Both missiles rushed at each other in a head-on engagement, the distance between them shrinking at hypersonic speed. At a predetermined distance, the Standard's proximity fuze triggered a symmetrical explosion, creating a wall of shrapnel the Chinese missile couldn't avoid. The M-19 shredded instantly. Its warhead detonated in the upper atmosphere, miles above the ocean and Navy ships it had targeted.

In the CIC, the chief petty officer turned and looked at the commander. "Standard-One intercept, sir! We got it!"

"Good shootin', people," the commander beamed. "Let Standard-Two run, just in case. Tell PACOM: All threats neutralized."

STRATCOM WARGAMING CENTER

"That's correct, Mr. Secretary," Aster said. "We believe Chinese subs are within an hour of achieving positions to threaten the closest battle group. We dodged Chairman Yi's first bullets by shooting down that swarm of four M-19 missiles. No ships were hit—but we were damned lucky, sir. This fight isn't over, though. Not by a long shot."

After the final M-19 had been destroyed, Aster had raced to the gaming center, and was sharing the Pit with Androsin. Aster had decided commanding the nation's global strike forces and space-based threat-warning assets from here made more sense, where he could monitor the real-time video link from Beijing.

News from the White House was disconcerting. The president's hold on STRATCOM's deterrent options was still in force, a frustrating, misguided strategy, in Aster's opinion. The hold order meant U.S. forces couldn't retaliate with overwhelming force, yet were subject to another attack from China.

Hit the bastards fast and hard! That had been Aster's combat philosophy for many years, and he believed in it wholeheartedly. Every minute that passed, while Boyer and the Washington crowd dithered, hoping for a diplomatic miracle, would cost lives. Aster bit his lip, repeatedly reminding himself that, in the United States of America, civilians controlled military forces, not the other way around.

The real wild card is Vandergrift. No way of knowing what damage that shit-bird has caused. What the hell was the idiot trying to do? Aster wondered, grimacing at the thought.

SecDef Hurlburt had relayed what little *he* had learned: Paul Vandergrift had been playing cat and mouse with Feng Bao Nu for more than four years. Motives were still unclear, but seemed to revolve around a warped effort to restore Vandergrift's tarnished reputation inside the Washington Beltway. The man had openly advocated a pacifist, appeasement strategy during the mid-nineties. Those ideas eventually had been integrated into the Boyer White House's approach to American-Chinese diplomacy, raising his stock significantly.

Vandergrift, believing Nu was a like-minded patriot—essentially an alter ego—within China's power elite, had cultivated a relationship that both men believed held great potential. In fact, the relationship was based on mutual, self-serving duplicity. Within influential U.S. circles, Nu had carefully nurtured sympathies for a strategy of U.S.-China engagement. Many had embraced the smooth-talking Chinese lobbyist's passionate vision. And Nu's considerable influence in Washington had paved the way for Vandergrift's eventual appointment as National Security Advisor. The new advisor had continued his relationship with Nu, convinced the two of them could reshape their nations and tip the scales of global power.

But once President Pierce Boyer confronted Vandergrift with the nightmare unleashed by the National Security Advisor's China gambit, the latter had confessed to his role. Still, he'd argued vehemently that his and Nu's clever strategy would still work to America's benefit. He urged Boyer to "Be strong! Don't waver now!" Trying to redeem himself, the man at least had the good sense to reveal what he knew would be occurring inside China's complex political machine.

"Howard, Van admits that Nu probably set him up, big-time," Hurlburt growled. "Van was naive, thinking he was doing—well, you know the ol' story: '*I and only I can save the world from horrible warmongers.*'"

However, Aster knew Vandergrift was no typical bleeding heart. He was a Machiavellian traitor, the bastard who had killed Stanton Lee.

"The hell with him!" Hurlburt spit. "Van claims Nu is really in cahoots with a bunch of hard-liners, a bunch the CIA's been tracking for some time. The hard-liners' strategy was to foster and nurture engagement by sucking up to us throughout the last decade, while China's economic might was growing. Of course, their military might was increasing, as well."

"But what about China's noise concerning 'cooperation'?" Aster countered. "You know, like the Pacific Rim trade initiative. Was that bullshit, as well?"

"That came from Yi, and it *was* valid, we believe. Looks like he at least still holds power over the *moderates*, but, in some ways, Yi could be a wild card from our perspective. We still need to drive a wedge of uncertainty smack through the Chinese leadership, and Plan Aurora is our best shot for doing that."

"It's underway, Mr. Secretary. Transitory information worms have been introduced to Aerospace City's control-center computers. Chinese comm and imagery satellites that were still operating after the North Korean EMP will soon go silent—as soon as the president gives us the go-ahead to pull the info ops trigger. China will go space-deaf and -blind, and stay that way, until we turn their birds back on. Say the word, and they're goners." Aster was pressing hard for action, and hoped the SecDef would do the same in the White House.

Hurlburt grunted, noncommittal, then changed subjects. "Sorry about Admiral Lee, Howard. We sure as hell lost a national treasure there."

"Roger that, sir. He knew what the risks were, but he went anyway. He tried to stave off the shit we're in now. One hell of a . . ." Aster's voice cracked, leaving a long silence. The SecDef signed off quietly.

Aster exhaled slowly, regaining his composure. *Damn! The ol' admiral's gone. Rare diplomatic skills backed by a warrior ethic— That's Lee's legacy.*

GREAT HALL/CHAIRMAN'S CONFERENCE CENTER

"We cannot assess damage to the American fleet? Preposterous!" railed China's PLA naval commander, his outburst of rapid-fire Mandarin resounding across the conference center. "Where is our post-strike reconnaissance? I *must* know how many ships we've sunk!"

No immediate answers were forthcoming, regardless of rants and demands. The electromagnetic pulse triggered by North Korea's nuclear detonation had played havoc with Chinese space-based intelligence assets, as well as those of the U.S. And the PLA's process of reallocating ground, sea, and air forces for intelligence, surveillance, and reconnaissance purposes was ponderous in the best of times. Now, in the heat of battle, it was gridlocked. China's Army had never developed the robust and agile military processes their counterparts in the U.S. had built over decades. An efficient, responsive structure for rapidly shifting assets and missions among commanders remained a mystery to PLA military district chiefs.

Chairman Yi rejoined the group of senior leaders. Some were seated around the conference table. Others hovered over workstations, demanding information from harried enlisted operators. Battle screens throughout the center displayed Chinese forces in a variety of readiness postures. One section of a large wall display showed a list of actions classed as imminent, those that could begin within the hour. Yi only had to touch his personal screen with a stylus to accept or reject the strategic options presented to him as a menu. He quickly reviewed them:

- *Destroy remainder of American carrier battle groups in China's Area of Responsibility.*
- *Launch missile and air salvos against Taiwan. Follow with amphibious assault per Sea Dragon plan.*
- *Release limited nuclear strike package against Guam (Andersen AFB) and Alaska's ballistic missile defense site (Fort Greely).*
- *Implement full retaliatory nuclear strikes against U.S. military, economic, and population centers.*

But Yi was reluctant, hesitant to act rashly. Pointing at the Knowledge Wall's third large screen, which displayed stark GloCon and *Aviation Week* accounts from Admiral Lee's China wargame, Yi again fired questions to those around him.

"Your retaliation options are the same as those predicted by the American wargamers! Are there no innovators in the vaunted PLA?

Look at the Americans' projected outcomes! Do you doubt that
Chinese cities will be attacked? *Nonsense!* Millions of our people
will be killed!"

The chairman's harsh questions and conclusions elicited only silence. Yi continued in another vein, "And why are the Americans
suddenly communicating with so many national command centers
around the world? Why the Indians? Why the Russians? Why *Vietnam*? I do not like this! There may be alliances in place we do not
fully understand. Those pacts could mean China will be attacked
from many directions! I will *not* act from ignorance! *Give me answers!*"

But there were no answers. The Chinese military leaders were
dumbfounded, confused by a flurry of U.S. communications detected
by the PLA's intelligence services shortly after the M-19 launches.
Much of the comm traffic appeared to involve strategic nuclear
forces, even though the content of U.S.-encrypted messages could
not be deciphered. It was the *pattern* of comm traffic that worried
PLA intelligence chiefs. No one could make sense of the American's
unexpected actions.

The Aurora Deception Plan was serving its purpose. Doubt and
confusion reigned in Beijing.

A terse report flashed on the conference center's displays, throwing the room's occupants into yet another state of loud confusion.

*Aerospace City reports loss of signals from Earth-observation,
telecommunications, and navigation satellites.* Yi was visibly
shaken and frustrated with his advisors' inability to make sense of
senseless data. A clamor near a conference center entrance drew his
and others' attention. Several men forced their way past PLA security guards, shouting, their words indistinguishable.

One voice caused Mi Soo to stifle an urge to scream in delight: her
father's. From the midst of the unruly crowd, Dr. Zhang emerged,
frantically searching the large room. Behind him, Feng Bao Nu was
shoved inside, causing him to stumble. Chairman Yi, taken aback
by Nu *not* being bound, wondered for an instant: Were his own security forces, the forces General Luan had ordered to find and arrest
Nu, allied with the renegade?

Without rising, Yi grabbed and rang a small bell resting innocuously near his personal computer display. The incongruous, Asian-culture tinkle quieted the room. Yi gestured the men to approach, while asking a technician to mute the video teleconference link. Not that it mattered, Mi Soo thought, glancing at the big wall-mounted screen. President Boyer had disappeared. The image from Washington showed only an empty chair and long conference-room table. A few people milled about aimlessly, but the president's distinctive figure was absent.

Dr. Zhang embraced his daughter warmly, tired features reflecting relief and grave concern.

Simultaneously, Feng Bao Nu drew himself up before Yi, defiant and brash. "Arresting *me*? Forcing *me* to come to *you*?" he retorted. "I demand . . ."

"*Enough!*" Yi barked, raising a hand, a leader's universal gesture demanding quiet. "You walk a fine line between patriot and traitor, Nu," Yi said softly, composed.

Nu sputtered, "And you, Chairman Yi—you—you have not explained your intentions! You are derelict . . ."

"Do not chastise *me*, Nu!" Yi shouted. "With great reluctance, I have acted on China's behalf, but act I have. A missile attack against the American battle groups was in response to their B-2 strike, a misguided attack you and the American traitor, Vandergrift, manufactured for your own gain! Your treachery now threatens the lives of *millions* of our people!"

"But missiles are not enough!" Nu cried, ignoring Yi's charges. "We must press our advantage! Destroying the capitalists' battle groups will force them to retreat. Bloody the Americans' noses, and they will withdraw in shame, just as they did in Iraq! Defeating America today will restore China's long-damaged soul tomorrow! While the U.S. is licking its wounds, Taiwan must be attacked! The Americans will *never* escalate, *never* risk nuclear war to save Taiwan!"

"Hmmmph," Yi waved dismissively. "You *think* you know a great deal about the Americans and their plans, what they will do if attacked. You are a fool! What about *your* plans!" Anger had crept into Yi's retort, his voice louder.

Zhang stepped away from Mi Soo, interjecting quickly, "I can answer that, Chairman Yi. Nu and his fellow traitors believe they can wrest control from you, *if* you fail to attack Taiwan. Each phase of this escalation with America is part of the cabal's strategy. It began with sacrificing the GBL site and its brave soldiers. It is part of a deal Nu made with the American National Security Advisor, Vandergrift."

Yi turned, open-mouthed and stunned. "Deal? What are you saying?"

"Do not be a fool, Chairman!" Nu screamed, lunging toward Yi. He was stopped by a burly Army guard's quick arm. "The Americans will bring the attack to us now! You have only minutes to order the next strike!" Nu turned to China's senior leaders, who had remained silent, stunned at the man's spectacle. Finding only hostile stares, Nu's composure slipped.

"Can't you see?" he shouted. "There is no time for talk! We must act! *Taiwan*! *Attack*! The Americans can no longer dictate to us! Vanquish the hegemons! *Vanquish them!*"

Murmurs and nodding gray heads indicated Nu's rambling vitriol was resonating with long-held sympathies. Mi Soo saw the Yi-Nu power struggle could tip either way in a heartbeat. Unobtrusively, she tapped a few keys at her position, reactivating the Washington-Beijing videoconference's audio link. Those in Washington could now hear everything being said in the Chinese center.

Mi Soo's high-pitched voice penetrated the rustle of murmurs. "Chairman Yi! We must contact the American President, *now*. I am no military strategist, but as a wargamer, I know the art of battle-and-consequence," she continued, pointing to the third screen, which listed STRATCOM's projected outcomes of a U.S.-China exchange. Heads turned her way.

"The American aircraft carriers were *never* a threat to China," she declared. "They were merely a response to North Korea's bluster and nuclear threat. The carriers' presence was turned to advantage by Feng Bao Nu's clever subterfuge. His plot is all about Taiwan; he wants to punish Taiwan severely. He wants history to record that he, Nu, is a Hero of the State for bringing Taiwan back into China's embrace, regardless of the international damage incurred by doing so."

"Preposterous!" Nu blustered, trying to wrest his arm free. The muscular security guard squeezed tighter, eliciting a painful yelp.

"No, not preposterous," Mi Soo countered. "My father discovered your plot to attack Taiwan. It has nothing to do with China's honor, and everything to do with *your* ambitions for power and riches. China's *true* leaders want unification, but not by killing Taiwanese and Chinese soldiers. Taiwan poses no threat other than ideology and pride. We cannot embark on such a senseless strategy!"

Dr. Zhang interrupted, his strong voice ringing. "Chairman Yi, my daughter is correct. This cabal's planning extends back to the Tiananmen incident. Nu and his traitorous comrades believed China's society was poised to unravel, and they needed a galvanizing issue to bind our people together. Taiwan became that center, a cause *they* believe would unite China through a cauldron of constant struggle. Just as Mao unleashed the horrors of the Cultural Revolution, a horror you experienced yourself, Mr. Chairman. Many of us know your own parents were killed by those mobs." He swung his arm to encompass the other leaders, some nodding.

"Taiwan would be the next struggle, and from armed victory would come a sense of greater glory, Nu and his gang postulated," he continued. "Our people would forget about their growing thirst for liberties, about misery in the Western provinces, and about the turmoil over Tibet. I opposed such imprudent thinking then, and I oppose it now. We *can* achieve peace and develop a better relationship with Taiwan—and with America, as well. But *not* by waging a fool's war under false pretenses! Reject this war, Mr. Chairman!"

"I, too, reject senseless war, Mr. Chairman!" The booming voice of President Pierce Boyer filled the Great Hall. Mi Soo cringed. She had turned up the audio link's volume far too loud.

"I was not aware of your return, President Boyer," Yi answered formally, regaining his composure. "Turn that volume down!" he added angrily. Facing Boyer's image at the far end of the conference table, Yi flashed a smile. *A smile of embarrassment*, Mi Soo thought. "How long have you been listening, Mr. President?"

"Long enough, Chairman Yi. I certainly agree with Dr. Zhang. And I pledge to work with you, regarding the concept of one China, if you will guarantee the political, cultural, and personal freedoms of Taiwan's people." He paused, letting that bombshell sink in.

Get the idea, Chairman? Boyer smirked to himself. A translator summoned to the White House Situation Room had arrived in time to decode the private debate being surreptitiously transmitted from Beijing. Right or wrong, eavesdropping had provided Boyer an invaluable negotiating edge. It might be exactly what he needed to stop this nonsensical escalation toward all-out war.

"Taiwan's people are not Americans," he pressed. "They are Chinese. But they are free people. Let them keep the freedoms they now enjoy. Let them live without fear of annihilation by their Chinese brothers. And let *their* freedoms take hold in your own great culture." It was bold, but Boyer was in his element, playing the world's leading statesman. He turned slightly, letting his eyes rest on the small woman off to one side.

"Mi Soo, I trust you have some acquaintance with the mobile command center Admiral Lee was using earlier?"

"Somewhat, Mr. President."

"Good. Please reestablish the communication and imagery link with the *Reagan* battle group. I want Chairman Yi to see the flag bridge's current condition, as well as its navigational fixes, and the course and speed the group is steering. Clear?"

"Loud and clear, Mr. President." Eavesdropping from Offutt AFB, General Aster silently added: *And maybe a little battle-damage-assessment imagery to show the good chairman that his missiles never dinged a single ship!*

WHITE HOUSE SITUATION ROOM

An aide handed a note to Secretary of Defense T. J. Hurlburt, who read it, smiled, and let his shoulders slump in relief. He nodded to Herb Stollach, the U.S. intelligence director, and flipped a thumbs-up to Gil Vega, the president's chief of staff. Vega closed his eyes briefly, then waved Hurlburt toward Boyer. The president was seated before a large screen dominated by the image of China's Chairman

Yi. The two leaders were locked in an intense exchange, something about Taiwan, the SecDef noted absently.

Hurlburt gently touched Boyer's arm, drawing his attention. He slipped the note in front of the president, then whispered to him. Boyer nodded, glanced at the note and leaned forward, facing the camera and display screen again. His demeanor changed from gentle pleading to strength and determination.

"Mr. Chairman, I am informed that our forces have intercepted all four of China's M-19 missiles that were fired at our carrier groups in the Pacific. The carriers are safe." He waited, not sure what to expect.

Yi stared into the screen, then glanced at the small computer screen before Mi Soo. It showed a high-angle satellite view of the *Reagan* carrier group, intact. His expression remained noncommittal for a few moments. Then he slowly smiled, nodding slightly.

"Mr. President, I am relieved, frankly. There has been enough bloodshed." His countenance clouded as a new thought crossed his eyes. "And what will be *your* response to those missile attacks, President Boyer? Will your wargamers' perceptive predictions come to pass?" He lifted a palm to the center's third screen, where the GloCon commentary and list of wargame outcomes were posted. A Chinese camera operator turned the lens toward that display and zoomed tighter, enabling those in Washington to read the text.

In the STRATCOM wargaming center, Aster shot a glance at Colonel Androsin, who was grinning like the proverbial feline with a sparrow in its mouth. *Maybe—just maybe it worked!* Aster hoped silently.

Behind him, surprise swept the center's wargamers, as they read the *Kosmos* GloCon's detailed interpretations of their long and tedious deliberations. None of them had been told that the China sidegame's classified outcomes were being released to the media. Only Aster, Androsin, the STRATCOM public affairs chief, and two reporters knew *why* those comments were now posted in a Chinese command center.

Come on, guys! Aster urged under his breath. *Stand down, for mankind's sake!*

Boyer stared at the screen. If shocked to see the detailed GloCon and *Aviation Week* commentaries about a classified American wargame, he gave no visible indication of it.

"Mr. Chairman, the United States will not—I repeat, will *not*—take further military action. I am commanding our forces to stand down. And you?"

Yi nodded slowly, then carefully leaned back in his chair, both hands on the table's surface, waiting for the camera to refocus on him. Yi seemed to shrink, as if a great load had been lifted.

"China, too, will stand down, Mr. President. I will take immediate action against Nu and those crying for American blood. And I am sure you, also, will—*reprimand* Mr. Vandergrift?"

Boyer smiled, eyes also revealing exhaustion. "Indeed, Mr. Chairman. Indeed I will."

Yi stared into the camera, then added quietly, "I trust we will speak again soon, Mr. President. On more friendly terms, yes?"

Boyer nodded and smiled. "Yes. Most certainly. I look forward to that, Mr. Chairman."